Centering Gender in the Era of Digital and Green Transition

Kristie Drucza • Amira Kaddour
Sujata Ganguly • Adel M. Sarea
Editors

Centering Gender in the Era of Digital and Green Transition

Intersectional Perspectives

palgrave
macmillan

Editors
Kristie Drucza
Includovate
Kampala, Uganda

Sujata Ganguly
Includovate
New Delhi, India

Amira Kaddour
National School of Advances Sciences and Technologies
University of Carthage
Amilcar, Tunisia

Includovate
Tunis, Tunisia

Adel M. Sarea
College of Business & Finance
Ahlia University
Manama, Bahrain

ISBN 978-3-031-38210-9 ISBN 978-3-031-38211-6 (eBook)
https://doi.org/10.1007/978-3-031-38211-6

This Palgrave Macmillan imprint is published by the registered company Springer Nature Switzerland AG.
The registered company address is: Gewerbestrasse 11, 6330 Cham, Switzerland

Paper in this product is recyclable.

Acknowledgment

The editors express their gratitude for the contribution and efforts of Mrs. Louise Hoole during the proofreading process.

Senior Editor, Includovate
louise.hoole@includovate.com
Website: www.louisehoole.com

Short biography: Louise was born in England, grew up on the remote Atlantic island of St Helena, and has spent many years in Africa. She is the author of seven books including *Talking to the Ancestors* (about East African art), Seven Wonders (about the world heritage sites of Tanzania), and Black Rock, a novel exploring the exile of Napoleon Bonaparte on St Helena. Louise travels widely between Africa, Asia, and Europe. She joined Includovate in 2023 as their senior editor. She is passionate about decolonization, whale sharks, and wild swimming.

Contents

Notes on Contributors

Alina F. Burlacu is a senior transport specialist with the World Bank Global Road Safety Facility, managing the Bloomberg Philanthropies Initiative for Global Road Safety (BIGRS) program, which is spreading across 15 countries. Starting from 2017, she has been spearheading road safety work with a diverse portfolio of Asia-Pacific governments and has been managing a diverse team of road safety professionals, working at the nexus of numerous national, municipal, and local governments throughout Asia-Pacific.

Mariette Correa Working for over three decades with national and international development and humanitarian organizations across the global south, Correa has focused her efforts on excluded and marginalized groups including people infected and affected by HIV/AIDS, girls and women who have been trafficked for prostitution, women facing violence, drug users, men who have sex with men, tribals, dalits/scheduled castes, small-scale fish workers, and Tibetan refugees. With a strong feminist orientation, she has built capacities of organizations; managed programs; conducted research, monitoring and evaluations; and developed organizational policies. Her PhD in Development Studies examined the contribution of women's studies in India to feminist methodology.

Natalia Pastori Curbelo is an evaluator and researcher with experience working on gender-related issues, such as gender-based violence, migrations and disability, international cooperation, and project planning and implementation within the public sector and NGOs. She has participated in a number of evaluations such as the Mid-Term Evaluation of the Sexual

Violence Research Initiative (SVRI) based in South Africa, the Outcome Evaluation of Light for the World's country programs in Bolivia and North-East India, and the Outcome Evaluation of Light for the World's "Equitable, Sustainable Eye Care for All" project in Burkina Faso, Mozambique, and Ethiopia.Natalia has effectively participated in collecting and analyzing raw qualitative and quantitative data; reviewing literature and creating comprehensive and rigorous documents on it; developing and implementing of research methodologies; writing reports to summarize findings and contribute to decision-making; developing and execution of monitoring tools; planning and implementation of projects related to gender violence and project coordination and implementation. She also worked on social network analysis mapping for projects such as Information and Studies Centre (CIESU).Natalia received her master's degree from the Complutense University of Madrid (UCM) in Government and Public Administration and from the Fluminense Federal University (UFF) in Social Policy, and a bachelor's degree from Universidad De La República (UDELAR) in Political science. She is a native speaker of Spanish, is fluent in English, and has proficient knowledge of French and Portuguese.

Kristie Drucza With over 17 years of experience as a community developer, gender and inclusion adviser, or researcher, Drucza is theoretically strong on gender, human rights/child rights, and social inclusion, particularly within a developing country context. She completed her PhD in Political Anthropology on Social Inclusion and Social Protection. She has a Master's in Applied Anthropology and Participatory Development specializing in Gender. Drucza has a passion for building partnerships and the capacity for systems change that leads to social inclusion. Systems change for social inclusion involves altering the underlying behavior, mental models, evidence, and relationships that keep the exclusion in place. It requires networking and peer-learning across power levels and structures. It is about changing the way people currently solve problems, and who is involved in examining these systems and structures so that exclusion is not reproduced. Those in the system have as much work to do as those outside the system.

Sinda Elghoul is an assistant professor at the University of Manouba. Her research interests include pattern recognition, image processing, computer vision, image recording and reconstruction in biomedical imaging and in cultural heritage. She currently works on topics related to machine learning, pattern discovery, classification, learning from big data, and learning in the context of a changing world.

Emna Fourati is an agronomist engineer specializing in fisheries and environmental engineering, graduated from the National Agronomic Institute of Tunisia. After a professional experience in the field of scientific research within the MARBEC in France: Marine Biodiversity Exploitation & Conservation whose supervisors are CNRS, UM, Ifremer, and IRD, she began her professional career in Tunisia in the field of climate policy. This experience started in 2017 with the organization RAJ-Tunisia as a project coordinator in its Environment and Climate department and continued with the Heinrich Böll Foundation as a climate change consultant to support its Environmental Policy and Sustainable Development department through the Local Climate Action project. During her professional experience, Emna focused on climate policy work at the international, regional, national, and local levels through launching initiatives, implementing projects, organizing conference debates, participating in COPs, attending stakeholder workshops, writing articles and guides, and producing videos and interviews. Thus, this professional experience was an opportunity to work on several themes related to change while involving different stakeholders.

Sujata Ganguly is a gender research specialist with more than seven years of experience. She did her MPhil and PhD in Population Studies from International Institute for Population Sciences, Mumbai, India. She has been intensively involved in gender research, communication research and monitoring, evaluation and learning activities in the field of migration, agriculture, and maternal and child health. In communication research, she did user tests of various mass media and mid-media campaign materials and evaluated its reach, recall and impact to inform the communications specific strategies. In gender research, integrated gender indicators in monitoring and evaluation frameworks, she developed tools to track and evaluate changes in women's empowerment. Ganguly's research publications include the changes in participation of women in rice value chains and its implications for control over decision making, guide to user testing communication materials, evaluation of India's soil health card from users' perspectives, husbands' out-migration and familial support for the left-behind wives from rural India.

Hela Ghbara is an assistant professor at the University of Tunis, Tunisia, specializing in the field of finance. With a strong academic background, she holds a PhD in Finance, which has equipped her with extensive knowledge and expertise in the financial sector. Her research interests revolve around various areas of finance, including investment analysis, portfolio management, and financial markets.

Imen Gmach is an assistant professor in Imam Abdulrahman Bin Faisal University, Dammam, KSA. She holds PhD in Management Sciences(management information system). Her works cover digital and inclusive society prospects.

Saida Gtifa is Assistant Professor of Finance at the University of Economic Sciences and Management of Nabeul in Tunisia. She has taught many courses in finance, accounting, and management, such as market finance, project management, financial analyses, corporate management, financial accounting, FOREX microstructure, and so on, in several Tunisian universities since 2005., She focused in her research on the following topics: Fintech, green finance, microstructure of foreign exchange market, stock market, and equality.

Kaleab Kebede Haile is a development economist who works as a principal researcher at Includovate. He has around 10 years of experience in development research and practice in the context of sub-Saharan Africa. His fields of research interest and expertise include the topic of climate resilience to understand, among other concepts, the nexus between climate change and women's economic empowerment, and how this intersects with the prevailing gender norms in the context of SSA. In particular, he is interested in assessing the role of national policies in eliminating or reducing the disproportionate physical and financial burden of climate change on women and girls related to increased transaction costs of accessing and utilizing clean cooking energy. His educational background includes a PhD in Economics and Governance from Maastricht University and a Master of Science in Agricultural Economics from Haramaya University.

Kanika Joshi is an international development practitioner and consultant. She comes with a multi-sectoral understanding of working with changemakers across the global south and is deeply passionate about youth building solutions for sustainable development. She works as the head of Partnerships and Communications at Includovate, a global feminist research organization, building thought partnerships around social inclusion and gender equality. Prior to this, she worked as a research manager with LEAD at Krea University in India on women entrepreneurship and financial inclusion for the last mile and in Kenya as a Fellow Consultant with TechnoServe on food security issues. She also contributes as a Global Senior Advisor to the Youth Solutions Program at UN SDSN Youth, where she has been key in setting up the Youth Solutions Report with a diverse team of passionate individuals.

Amira Kaddour is an associate professor at the Higher School of Sciences and Advanced Technologies, Carthage University. She is the chair of the Professional Master's program: Innovation Engineering and Technology Transfer and the director of the center for training and certification of the school. Since September 2008, she served Tunisian universities for different subjects linked to social and economic sciences. She received her PhD in Management and Finance-Sciences from Manouba University. She has professional certifications in FinTech from HKU. Currently, she is focusing on digital and green transition in her works. She is the president of the International Society for Knowledge Organization-Tunisian chapter. She is an active member in the national civil society, African CSOs, and a member in the executive board of the African Women in FinTech and Payment. Her research interest spans financial inclusion, gender equality, intersectionality and gender transformative approach, SDGs in low- and medium-income countries, inclusive society, behavioral finance, FinTech, and green finance. Much of her work has been on improving the understanding of risk-taking behavior in the case of financial turmoil, mainly through the application of event study analysis, econometric models, deep learning through wavelet neuronal networks, in addition to the use of advanced technologies in finance to explore investor's behavior. KADDOUR has contributed to the steering committee of several conferences and projects, focusing on inclusive economy, vulnerability, and outlooks of green and digital transition, in addition to study on gender transformative approach in climate change and water security in Africa, STEM, and STEAM education and gender-based violence in Tunisia for different projects. She has contributed to different projects toward efficient implementation of SDGs and UN agenda for road safety 2020–2030. She has given numerous talks and is editor of many books.

Nadia Labidi is Assistant Professor of Finance at the Higher School of Communication of Tunis. She taught several subjects in finance, accounting and management in Tunisian Universities for 20 years. She participated in many projects in the national and international context in project management and entrepreneurship as a tutor, coach, trainer and member of organization committee. Her research is focused on corporate finance, Islamic finance, Fintech, green finance, entrepreneurship, and gender equality.

David Lefor is an educational expert with more than 10 years of experience in teaching demographic and social issues in developing economies. He has extensive experience in the Middle East and South East Asia and has conducted in-depth research into the impacts of statelessness and refugee status on populations in Myanmar and Thailand. He has been involved in gender research and monitoring, evaluation and learning activities with a wide range of NGOs and intergovernmental organizations. He holds an MSc International Development from the University of Portsmouth, a BEd from Lakehead University, and a BA from Carleton University.

Sinda Lghoul is an assistant professor at Manouba University Tunisia.

Maissa Louhichi is an agronomist engineer specializing in fisheries and environmental engineering. She graduated from the National Agronomic Institute of Tunisia. She also holds Research Master's Degree in Operations and Management of Aquatic Ecosystems and Research Master's Degree in Statistics and Data Analysis Applied to Health Sciences. In her PhD thesis she worked on endangered species and proposing mitigation measures to reduce by-catch of different species. She has an experience in conducting international research projects focusing on vulnerable species. Maissa holds a professional master's degree in innovation engineering and technology transfer from National School of Advanced Sciences and Technologies of Borj Cedria, Tunisia.

Amira Maazouz is currently Assistant Lecturer and Researcher of Law at the Faculty of Legal, Political and Social Sciences of Tunis, University of Carthage. She holds Master's in Common Law and PhD in Private Law from the Faculty of Legal, Political and Social Sciences of Tunis, and she is also a lawyer who is registered at the Tunisian Bar Association since 2010. She taught at many Tunisian Universities since 2011, and she served as head of Department of law from 2017–2022, head of Professional Master's in Business and Digital Law from 2019–2022, and Master's in Judicial Law from 2020–2022 at Law and Business School (Tunisian

University). She is a member of Research Lab LR-DIERME known as *Laboratoire de recherche en droit international et relations Maghreb-Europe*, which was founded in 2014 and is part of the Faculty of Legal, Political and Social Sciences of Tunis, University of Carthage. Her teaching and research interests include common law, international commercial law, international investment law, arbitration law, financial markets law, company law, contract law, IT law, and RegTech.

Emmanuel Kodwo Mensah is an associate researcher at Includovate. He has worked in various organizations. He completed Master's in Gender Studies from Makerere University, Uganda. He also has Bachelor of Sciences in Community Nutrition from University of Development Studies, Ghana. He is team player with a profound impact on the team and achievements of project goals. He has hands-on experiences in designing gender-responsive projects and MEL plans. He is an expert in gender, climate change, agriculture, food and nutrition security (CCAFS), financial and digital inclusion for rural development, youths and women empowerment, and positive masculinity to ensure male's accountability for gender equality. He has also investigated intra-household decision-making for Adoption of Climate-Smart Agricultural Practices in Nwoya District, Northern Uganda, and contributed a book chapter to the CGIAR Research Program on Climate Change, Agriculture and Food Security's Partnership for Scaling CSA (P4S).

Elena Nikolova is Includovate's Quantitative Research Director. She has more than 18 years of experience in economic and social development acquired in international organizations, academia and the private sector. She has worked on Eastern Europe, Asia, and Africa and has published widely on social and economic development, entrepreneurship, gender, well-being, political economy, and governance, specializing in quantitative and survey analyses. She has consulted for the Asian Infrastructure Investment Bank, the World Bank, the IMF—Africa Training Institute, the African Development Bank, the European Commission, several UN agencies, and the Council of Europe Development Bank. She is an expert for the European Institute for Gender Equality and FAO. She spent seven years at the European Bank for Reconstruction and Development (as a principal economist) and also was Assistant Professor in Economics at UCL. She is affiliated with several universities in Europe and is a lecturer at EBS Business School in Germany. She holds a PhD in Public Policy from Princeton University and is fluent in English, Bulgarian, Russian, and Italian. See https://sites.google.com/view/elena-nikolova

Adel M. Sarea is an associate professor and the *winner of Emerald Literati Award 2014* (UK). He possesses over 100 publications indexed in [SCOPUS, SSCI, ESCI, ABDC, ABS], and he has more than a decade of teaching experience. He has been bestowed with many awards and been summoned as the keynote speaker on several occasions. He serves on the editorial team of several international journals published by Elsevier, Emerald, and Taylor & Francis. He is a member of the Accounting and Finance Association of Australia and New Zealand (AFAANZ). In 2015, Sarea has been appointed as a recognized *Ph.D. supervisor by Brunel University London*. In 2021, he has been entitled as the senior editor of the *International Journal of Emerging Markets* at Emerald Publishing (UK); associate editor, *Heliyon Business and Economics* (Elsevier); consulting editor, *Journal of Education for Business* (Taylor & Francis); associate editor, *International Journal of Islamic and Middle Eastern Finance and Management* (Emerald); associate editor, *ISRA International Journal of Islamic Finance* (Emerald); associate editor, *Asian Journal of Accounting Research* (Emerald); associate editor, *Journal of Islamic Accounting and Business Research* (Emerald); guest editor, *Journal of Financial Reporting and Accounting* (Emerald); and associate editor, *MENA Journal of Cross-Cultural Management-Inderscience* (UK).

Jasmina Bunevska Talevska received her PhD from the University of Bitola, Faculty of Technical Sciences Bitola, Department of Traffic and Transport, and Post-Doctoral specialization in Road Traffic Safety in 2020 from the European Transport Safety Council and University of Hasselt Summer School. Her teaching-educational experience has covered all the necessary degrees, continuously advancing through many years of engagement in the fields of "Traffic Engineering and Design," "Traffic Design of Road Infrastructure," and "Road Safety and Sustainable Urban Mobility." In her university career and theoretical and applicative scientific research, as an author or co-author, she has published significant number of papers. She gained experience through participation in expert committees and bodies; work on scientific research, application, and expert projects; membership in domestic and international professional associations; presenting an impressive number of international conferences and congresses; and visiting prestigious European universities as a guest lecturer. She strengthened her professional development through ERASMUS and CEEPUS scholarships for teaching. She is a *winner* of the photo competition launched as part of the European Day without a Road Death campaign in 2018 from the European Commission, Directorate General for Mobility and Transport. In 2019, she received an Honorary Title "Ambassador of Traffic Culture" from Hungarian Association of Transport Sciences, and in the same year, she won the prestigious municipal award 4th of November AWARD of the city of Bitola for the achievements in the education and technical culture for 2019.

List of Abbreviation

ADBI	Asian Development Bank Institute
AIC	Artificial Intelligence Community
CBT	Climate Budget Tagging
CEO	Chief Executive Officer
CGIAR	The Consortium of International Agricultural Research Centers
COP	Conference of Parties
CPCS	Canadian Pacific Consulting Services
DEI	Diversity, Equity & Inclusion
EC	European Commission
ECOWAS	Economic Community of West African States
EU	European Union
GAD	Gender and Development
GAP	Gender Action Plan
GBV	Gender-based Violence
GDP	Gross Domestic Product
GESI	Gender Equality and Social Inclusion
GGGI	Global Green Growth Institute
GHG	Greenhouse Gas
GIP	Green Industrial Performance
GSMA	Global System for Mobile Communications Association
GTF	Garment, Textile and Footwear
IAEG-GS	United Nations Inter-Agency and Expert Group on Gender Statistics
ICT	Information and Communications Technology
IPCC	Intergovernmental Panel on Climate Change
IRB	Institutional Review Board
ITU	International Telecommunication Union

LGBTQIA+	Gay, Lesbian, Bisexual, Transgender, Queer, Intersex, and Asexual people collectively
LMICs	Low- and Middle-Income Countries
MOOCs	Massive Open Online Courses
NDC	Nationally Determined Contributions
NGO	Non-Governmental Organization
NIDL	New International Division of Labor
OECD	Organization for Economic Cooperation and Development
OHCHR	Office of the High Commissioner for Human Rights
RMG	Ready-made Garment
SDGs	Sustainable Development Goals
SMEs	Small and Medium-sized Enterprises
STEM	Science, Technology, Engineering, and Mathematics
STRIA	Strategic Transport Research and Innovation Agenda
TinnGO	Transport Innovation Gender Observatory
TRIMIS	Transport Research and Innovation Monitoring and Information System
UN	United Nation
UNCTAD	United Nations Conference on Trade and Development
UNEP	United Nations Environment Programme
UNFCCC	United Nations Framework Convention on Climate Change
UNGC	United Nations Global Compact
UNIDO	United Nations Industrial Development Organization
UNIFEM	United Nations Development Fund for Women
WDI	World Development Indicators
WEE	Women's Economic Empowerment
WEF	World Economic Forum
WHO	World Health Organization
WID	Women in Development

List of Figures

List of Tables

List of Boxes

CHAPTER 1

Introductory Chapter. Gendered Challenges in the Digital and Green Transition: An Intersectional Exploration

Kristie Drucza and Amira Kaddour

1.1 Overview

Addressing the challenges of scarcity of natural resources, poverty, and disparities in the twenty-first century requires a focus on implementing economic development based on digital and green transition, with a strong emphasis on social welfare as a fundamental and sustainable goal. While current economic models have led to growth, inequalities persist. These models have not fully resolved the root causes of hunger and

K. Drucza
Includovate, Kampala, Uganda
e-mail: kristie.drucza@includovate.com

A. Kaddour (✉)
National School of Advances Sciences and Technologies, University of Carthage, Amilcar, Tunisia

Includovate, Tunis, Tunisia
e-mail: Amira.kaddour@ensta.u-carthage.tn; amira.kaddour@includovate.com

K. Drucza et al. (eds.), *Centering Gender in the Era of Digital and Green Transition*,
https://doi.org/10.1007/978-3-031-38211-6_1

malnourishment and have resulted in disconnects between what is needed for people and the planet to thrive, and what profits for companies are based on. These models caused severe environmental damage and contributed to climate change. Interconnected injustices exacerbate political divisiveness and popular unrest they lead to conflicts such as escalating resource scarcity and uncontrolled migration, destabilizing economies, and threatening social cohesion. The worst impacts of these interconnected injustices are felt by those who are subjected to systemic discrimination, particularly based on gender and race.

As we look toward the Mega trends of the century, it becomes clear that we are transitioning to a new model of life, heavily influenced by advanced digital technologies such as artificial intelligence. This transition will bring about continual change in past and present modes of communication, with massive disruptive innovation on one hand, but also an important need for a universal language of inclusion, rather than exclusion, on the other. To achieve sustainable development and societal progress, investing in the human factor is key. Education plays a crucial role in this by preparing future generations with the necessary skills and knowledge to master the basic sciences, while also instilling ethical, civic, and environmental responsibility in the face of the limitless possibilities offered by artificial intelligence. To promote diversity and economic inclusion, education must also emphasize these values. The past crisis of COVID-19 highlights the need to adapt nations to meet unconventional and unexpected needs.

1.2 Current and Future Challenges

The Sustainable Development Goals for the new decade of 2021–2030 aim to guide nations and the global community toward an inclusive society through a series of ambitious targets. These goals are driven by the need for transformative change, including the integration of advanced technologies and the ability to withstand challenges such as natural resource scarcity and climate change.

At this level UN Women in their study *The Digital Revolution: Implications for Gender Equality and Women's Rights 25 Years After Beijing (2020)* stipulates that "the digital revolution brings immense potential to improve social and economic outcomes for women. Yet, it also poses the risk of perpetuating existing patterns of gender inequality. Despite several important initiatives, a significant digital gender gap remains, limiting the equitable realization of the benefits of digital transformation across high-, low-, and middle-income countries" (Wajcman et al. 2020). However, the emphasis on resource efficiency and the shift toward environmentally friendly solutions are closely tied to the utilization of technology, indicating that the promotion of digital transformation is a crucial element in the successful implementation of the green transition. The European Commission (2022) stipulates that "during the Digital Decade, Europe faces two important challenges: the green transition and the digital transition. These might seem like two distinct issues, but really, they are twin challenges: neither can succeed without the other. And, they are both equally important for Europe's future" (EU 2022).

1.3 The Book's Contribution

This book aims to inform readers, including academicians, researchers, and policymakers, about the significance of gender in the current challenges of the digital and green transition. It aims to provide insights to improve socioeconomic programs and policies toward inclusive development. Through various chapters, the book presents the relevance of considering the gender dimension in major socioeconomic aspects of digital and green transformation in a transversal way, with the objective of guiding the reader toward the recognition of the need for gender transformative approaches. It covers basic dimensions, such as the production and analysis of disaggregated data, and extends to a sectoral vision, highlighting the relevance of the gender aspect in the acceleration and equitable success of these twin transformations. Additionally, the book aims to fulfill educational needs by highlighting the aspect of intersectionality when analyzing the impact of the gender factor in the digital and green transition and in the achievement of the Sustainable Development Goals (2030). The book is divided into four complementary parts: the first part raises awareness of the importance of the issue in the context of current global challenges, the second part proposes a sectoral analysis, the third part addresses cultural challenges and the need to strengthen the role of women, and the last part nuances the positive impact of these considerations on economic growth and poverty eradication. A technical tool of great importance is presented in the last part, which allows understanding, informing, and monitoring of a just transition using gender-sensitive data.

References

Shaping Europe's digital future. 2022. Green digital sector Europe's digital transition goes hand in hand with the European Green Deal. https://digital-strategy.ec.europa.eu/en/policies/green-digital Feb 2022.

Judy Wajcman, Erin Young, and Anna Fitz Maurice. 2020. The digital revolution: Implications for gender equality and women's rights 25 years after Beijing. UN Women discussion paper series

PART I

Global Challenges

CHAPTER 2

Gender Within the SDGs Agenda: Importance and Intersectionality

Kanika Joshi, Emmanuel Kodwo Mensah, and Mariette Correa

2.1 Introduction

Gender equality is critical to progressing all 17 of the United Nations' Sustainable Development Goals (SDGs); it is also a human rights issue. This chapter maps the relevance of gender to four specific SDGs, two of which relate to digital transition and two of which relate to green transition. It starts by considering the gender dimensions of Goal 8 ("promoting inclusive and sustainable economic growth, employment and decent work") and Goal 9 ("building resilient infrastructure, promoting sustainable industrialization and fostering innovation"), and analyzes how these can impact the digital transition. It goes on to consider the gender

K. Joshi (✉)
Open Data Institute, London, UK
e-mail: kanika.joshi@theodi.org

E. K. Mensah • M. Correa
Includovate, Queanbeyan East, NSW, Australia
e-mail: emmanuel.mensah@includovate.com

K. Drucza et al. (eds.), *Centering Gender in the Era of Digital and Green Transition*,
https://doi.org/10.1007/978-3-031-38211-6_2

dimensions of Goal 12 ("ensuring sustainable consumption and production patterns") and Goal 13 ("taking urgent action to combat climate change"), and analyzes how these may impact the green transition.

Gender equality does not entail transferring opportunities from men to women, but creating conditions whereby every person—regardless of their gender—has the right and ability to realize their human potential (UNICEF and UN Women 2013). The structural, political, and representational barriers that "leave people behind" and the way these can be reproduced with unequal solutions will be discussed with reference to India, Ethiopia, and Australia.

The chapter recognizes the significant lack of gendered data on environment issues and the frequent lack of gender considerations in dialogue, planning and action on the environment. It identifies gaps in the SDG indicators, explores how gender is (inadequately) integrated, and argues that without adequate emphasis and incorporation there, we risk impeding overall progress to gender equality. The chapter further recognizes men and women's different needs, contributions, and impacts in relation to the green and digital transition, and considers how women's economic empowerment and political representation can drive the gender equality discussion forward. It argues that intersectionality is a necessary lens through which to understand and achieve gender equality and empower all women and girls (the specific aim of Goal 5). Moving beyond gender, it briefly examines other positionalities, such as persons with disability, recognizing there is even less data on these and other groups who are not adequately considered in the SDGs. The chapter ends with recommendations for strengthening the SDG indicators and overcoming the structural barriers that impede gender equality in the green and digital transition.

2.2 Methodology

Gender equality lies at the heart of the Sustainable Development Goals, cuts across all 17 SDGs, and is embedded in 45 targets and 54 indicators. This section outlines the methodology used by the authors to consider how gender is integrated in the SDG indicators that relate most closely to the green and digital transitions.

The authors first examined all the SDG goals and targets in order to identify those that related to green or digital transitions; targets were then narrowed down to Goals 8, 9, 12, and 13. Within these, the authors identified gender-disaggregated indicators to explore in more detail. The authors also looked at non-gendered indicators to assess the extent to which intersecting "vulnerabilities" were addressed. Goal 5 was included as a reference to determine how "achieving gender equality and empowering all women and girls" can be linked to Goals 8, 9, 12, and 13 with regard to climate change and the digital transition.

Once indicators had been identified, the United Nations' SDG Indicators: Metadata repository (2021) was used to determine which indicators had gender-disaggregated data; the concepts and definitions of each indicator were then assessed to ensure their relevance to the specific focus of the authors. Additional information and context were gleaned from the websites of various UN agencies, international organizations, and SDG tracking sites[1] (see Roser and Ortiz-Ospina 2018; UN Women n.d.-a, n.d.-b, n.d.-c, n.d.-d, n.d.-e), with a focus on three select countries (India, Ethiopia, and Australia). Insights from the selected indicators, plus the available gender-disaggregated data from focus countries, were used to provide relevant recommendations.

Given the paucity of data on the green and digital transition, as well as on our three focus countries, our next step was to collect data, where possible, from national government sources in order to supplement global data sets. This entailed accessing the websites of relevant government ministries and departments, as well as any key reports that shone a further light on these indicators. The inadequacy of data from UN, international, and SDG tracking sites, as well as the need to explore non-governmental viewpoints, led the authors to examine peer-reviewed and gray literature (both global papers and those specific to the three countries). These gave a civil society perspective to supplement our academic analyses of the SDG targets and indicators, and strengthen our recommendations.

2.3 Analysis of Indicators

This section provides a comparison between Ethiopia, India, and Australia (considered here as two developing countries and one developed country, respectively) and attempts to show how gender—within the broader context of intersectionality—maps to key indicators and targets for Goals 8, 9, 12, and 13.

[1] UN Women's SDG Indicator Dashboard provides information on the progress of SDG targets and indicators from a gendered perspective. Available at https://data.unwomen.org/data-portal/sdg

Gender and Climate Change (Goals 12 and 13)

Using an intersectional lens to address issues of climate change allows us to highlight how different groups of people are affected according to their specific contexts and social locations. The threat of climate change is increasingly acknowledged in political agendas, but issues of intersectionality and equity are often missing in climate change debates (Terry 2009). Intersectionality is an essential analytical tool to address the complexity of climate injustice. The sociopolitical locations of individuals and groups inform their participation, responsibilities, and vulnerability with regard to climate change. One such critical social structure, which the authors focus on here, is gender.

It is well recognized that climate change has different impacts on women and men, and that the former are more vulnerable to climate change impacts. Various studies have highlighted that climate injustice is gendered, and that women and men respond to—and are impacted by—climate change in different ways (OECD 2021). Social, political, and economic barriers impede the capacity of women, and these—combined with unequal access to resources, lack of decision-making power, a lack of mobility, and a high dependence on natural resources—increase women's vulnerability (WomenWatch 2009). Women have taken the lead in climate action in several countries, although their impact in these initiatives is more assumed than evidence-based (Price 2018). Acknowledging that women are more impacted by climate change than men, the Office of the United Nations High Commissioner for Human Rights (OHCHR) has taken steps toward gender-responsive climate action (OHCHR n.d.). The Fifth Assessment of the Intergovernmental Panel on Climate Change, for example, made intersectionality explicit in its Summary for Policymakers (IPCC 2014, 6):

> *Differences in vulnerability and exposure arise from non-climatic factors and from multidimensional inequalities often produced by uneven development processes (very high confidence). These differences shape differential risks from climate change … People who are socially, economically, culturally, politically, institutionally, or otherwise marginalized are especially vulnerable to climate change and also to some adaptation and mitigation responses (medium evidence, high agreement). This heightened vulnerability is rarely due to a single cause. Rather, it is the product of intersecting social processes that result in inequalities in socioeconomic status and income, as well as in exposure. Such*

social processes include, for example, discrimination on the basis of gender, class, ethnicity, age, and (dis)ability.

In December 2015, parties to the United Nations Framework Convention on Climate Change (UNFCCC) reached a landmark agreement to combat climate change and to accelerate and intensify the actions and investments needed for a sustainable low-carbon future. Commonly referred to as the Paris Agreement, it sets out detailed actions to help achieve the ambitious target of keeping global temperatures rise "well below 2°C above pre-industrial levels" and limiting global warming to 1.5 °C by the end of this century.

Countries (especially the most vulnerable) will need help to achieve these objectives including appropriate financing, technologies, and capacity-building initiatives. Signatories of the Paris Agreement are expected to take concerted action, and to report periodically on their emissions as well as their efforts to keep these under control. Australia, India, and Ethiopia are all signatories to the Paris Agreement. The preamble to the 2015 Paris Agreement notes that "Parties should, when taking action to address climate change, respect, promote and consider their respective obligations on human rights, the right to health, the rights of indigenous peoples, local communities, migrants, children, persons with disabilities and people in vulnerable situations and the right to development, as well as gender equality, empowerment of women and intergenerational equity." Parties also agreed to a five-year Enhanced Lima Work Program on Gender and a gender action plan.

Each signatory of the Paris Agreement is expected to report their Nationally Determined Contributions (NDC), which quantify carbon emissions reductions and adaptations to climate change. The final synthesis report of the UNFCCC secretariat, published in September 2021, revealed that most parties linked their NDCs to their commitments to transition to a low-carbon, resilient economy, as well as SDG goals, targets, and indicators. The report showed that more state parties now reference gender integration and include gender-related information in their new or updated NDCs. Furthermore, many state parties now consider gender as a cross-cutting issue in climate adaptation and mitigation, and elaborate on gender aspects in specific sectors (including energy, agriculture, health, disaster, water, land use and forestry, fisheries, and education). Some parties have also recognized the vulnerabilities, rights, roles, and situation of indigenous people, and acknowledged the importance of local knowledge to strengthen climate efforts.

Looking at the NDCs of our three focus countries—Australia, India, and Ethiopia—we find that Australia and India's NDCs do not have any information on gender (Australian Government 2021a, b; Government of India 2016). Australia did involve communities, as well as environmental and indigenous groups, in their review of their climate change policy of 2017. Australia's updated NDCs also refer to the planning processes that were used to prepare their NDCs, which included public participation and engagement with local communities and indigenous peoples in a gender-responsive manner (Australian Government 2021a, b). Ethiopia's updated submission acknowledges that, in order to achieve equitable outcomes, climate action should consider gender issues, and include vulnerable groups in design and implementation (Federal Democratic Republic of Ethiopia 2021). However, this statement is not elaborated further in the document.

Absence of Gender-Inclusive Data

Despite the disproportionate impact of environmental issues on women and their lack of participation in decision-making, the SDG targets fail to recognize the importance of gender in relation to environmental issues (UN Women 2019). The SDGs that relate to climate action (goal 13), promote sustainable consumption, and protect terrestrial (15) and aquatic ecosystems (14) include just one gender-specific indicator, 13.b.1. (UN Women 2016)[2] across the four environment-related goals. Unfortunately, there is no available data on this indicator from Australia (UN Women n.d.-a, n.d.-b, n.d.-c, n.d.-d, n.d.-e), Ethiopia (UN Women n.d.-a, n.d.-b, n.d.-c, n.d.-d, n.d.-e), or India (UN Women n.d.-a, n.d.-b, n.d.-c, n.d.-d, n.d.-e).

The SDG metadata repository template states in relation to the NDCs: "Starting in 2023 and then every five years, governments will take stock of the implementation of the Agreement to assess the collective progress

[2] UNSD, Sustainable Development Indicator 13.b.1: "Number of least developed countries and small island developing States that are receiving specialized support, and amount of support, including finance, technology and capacity-building, for mechanisms for raising capacities for effective climate change-related planning and management, including focusing on women, youth and local and marginalized communities."

towards achieving the purpose of the Agreement and its long-term goals" (United Nations Statistics Division (UNSD) 2021). The repository also references the National Adaptation Plan process, established under the Cancun Adaptation Framework, which is intended to be "a continuous, progressive and iterative process which follows a country-driven, gender-sensitive, participatory and fully transparent approach supported by technical guidelines" (UNSD 2021). Despite these aims, plans are currently unavailable for Australia, Ethiopia, and India.

Australia does, however, have a National Climate Resilience and Adaptation Strategy, which aims (as one of its two guiding principles) to support those who are vulnerable to climate-related impacts, to respect the knowledge and experience of those affected, and to involve them in decision-making (Australian Government 2021a, b). Australia also funds adaptation programs that are gender-responsive and socially inclusive in order to strengthen community resilience in partner countries (including Uganda, Zimbabwe, and the Philippines), especially for the most vulnerable.

Ethiopia's National Adaptation Plan also refers to building social protection and livelihood options for vulnerable people, with a special emphasis on women, children, and impoverished communities (Federal Democratic Republic of Ethiopia 2019). It also recognizes that women may be particularly vulnerable to climate impacts due to socio-economic inequalities that limit their adaptive capacity, and recognizes their short-term or seasonal coping mechanisms. The plan's guiding principles also recognize the importance of gender sensitivity, acknowledging that women must be given due attention because they are often more vulnerable to climate change impacts. Ethiopia's plan recognizes that climate change affects different groups in different ways, and that the participation of diverse groups in solutions is therefore essential. It "aims to integrate actions that improve the status of women and the welfare of children" and to ensure climate change measures are "planned and implemented in a manner that addresses the wellbeing of the elderly, persons with disabilities and environmental refugees." In addition, Ethiopia's communications strategy for adaptation targets women's groups (among others). However, there is paucity of gender-disaggregated data to measure achievements toward the SDGs.

The authors examined three specific national sources for target 13.b. indicators and found a similar paucity of data. The Australian government states in regard to indicator 13.b.1 that "the globally agreed methodology has not been set for this UN SDG Indicator, therefore Australia has not yet investigated potential data sources" (Australian Government 2018). In India, although there have been isolated efforts at developing gender-just climate solutions (Sami et al. 2016; Women and Gender Constituency 2017), there is no comprehensive gender-inclusive climate action. The SDG India Index of 2020–2021 states that India has identified five national-level indicators to assess its performance under SDG Goal 13 (NITI Aayog 2021) but none is gender-specific. In 2011, an African Development Bank economic briefing document analyzed climate change impacts on women working in certain occupations (primarily farming) and women's adaptation to environmental realities, and highlighted the need for gender analysis to underpin the development of climate change policies. While the briefing pre-dated the formulation of the SDGs and no gender-disaggregated data on climate change was available, the brief set out to assess "how best to ensure that robust data are made available for policymakers and advocates for gender equality and women's empowerment" (Nabalamba et al. 2019).

Data is crucial for policymakers. When there is only a single gender-specific indicator across the entire "gender–environment nexus" (UN Women 2019)—and when even that data is not gender-disaggregated—vulnerable groups and genders and their specific concerns are invisible to policymakers. The lack of gender-disaggregated data in environmental issues is an impediment to formulating and implementing effective policies on gender and climate injustice. Gender equality and gender-sensitive data collection and analysis must be considered as critical and urgent for the SDG agenda of every country (Seck and Maskey 2019; Encarnacion and Maskey 2021). In an effort to reduce gender data gaps in environmental policies, the OECD has introduced three new indicators (OECD n.d.):

- Mortality rates from air pollution, differentiated by pollutant, sex, country, year, and age
- Development of green technologies, differentiated by domain, sex, country, and year
- Exposure to environmental risks, differentiated by pollutant and river flooding, sex, and age.

In its report *Gender and the Environment*, OECD also carried out detailed analysis of the links between gender equality and environmental sustainability, looking at the nine environment-related SDGs (2, 6, 7, 9, 11, 12, 13, 14, and 15). This report recognizes the lack of a gender dimension to environmental issues and the considerable lack of data on the "gender-environment nexus" (OECD 2021, 18).

Gender and Responsible Production and Consumption

Efforts to foster sustainable consumption and production tend to focus on promoting efficient resource/energy use, sustainable infrastructure, and green jobs (in line with Goals 12 and 13). Research shows that women play a pivotal role in natural resource management, and that their increased participation leads not only improves local natural resource governance, but increases conservation efforts and sustainable livelihoods. Women tend to make the majority of consumer decisions and are more likely to be sustainable consumers, having a higher inclination to recycle, save energy, and reduce waste. Ensuring more women are in senior positions (in governments, development agencies, planning departments, and NGOs) fosters gender-sensitive approaches and initiatives in environmental management and—in turn—the more efficient use of natural resources (World Bank 2012).

Despite all this, not a single target under Goal 12 has gender-disaggregated information, though women typically have little or no access/control over productive resources (such as land, livestock, and agricultural inputs), as well as more limited access to financial services, technologies, training, information and marketing (Fletschner and Kenney 2014).

Gender and the Digital Transition (Goals 8 and 9)

SDGs: The Implications of Disaggregating Data by Sex and ICT-Related Factors

The United Nations Inter-Agency and Expert Group on Gender Statistics (IAEG-GS) identifies only two gender-relevant indicators pertinent to SDG 5 on digital transition. Target 5.b. measures "enhanced use of enabling technology, in particular information and communications technology, to promote the empowerment of women"; while 5.b.1 measures the "Proportion of individuals who own a mobile telephone, by sex." According to the International Telecommunication Union (ITU) and

IAEG-GS, this indicator is intended to monitor and improve gender equality through the design of targeted policies and action plans to address gendered digital divides, especially in developing countries. UN Women's *Progress on the Sustainable Development Goals: The Gender Snapshot* 2021 report emphasizes the prevailing gender divide between men and women's phone ownership (UN Women 2021). Thus, targeted policies are needed to close the gender gap in mobile telephony and ensure progress toward SDG target 5.b. (ibid.).

An examination—using the SDG tracker—of target 5.b.1. (the "proportion of individuals who own a mobile telephone, by sex") shows no sex-disaggregated data for Ethiopia and no data at all for India or Australia (Roser and Ortiz-Ospina 2018). There are also no data for this indicator in the SDG Global Database which is maintained by the United Nations Department of Economic and Social Affairs (United Nations Department of Economic and Social Affairs 2022). While Ethiopia has data for indicator 5.b.1 from 2015 and 2016, it is not gender-disaggregated (Roser and Ortiz-Ospina 2018). From a gender perspective therefore appropriate data does not exist to enable monitoring of this area.

Implicit in target 5.b.1. is the idea that mobile phone ownership and the empowerment of women are directly linked (see Porter et al. 2020). However, the structural and underlying causes and social norms that influence and perpetuate gender inequality are often not considered. For example, mobile phone ownership is only disaggregated by sex without recognizing class, disability, or poverty level, the different experiences and needs of specific groups, or the intersecting inequalities and structural power relations that prevent the realization of full inclusion and equity. This is particularly so in the context of LMICs—such as Ethiopia and India—where poverty is further impacted by gender, (dis)ability, class, and rural/urban location (Blackden 1999; Sender and Smith 2012; USAID 2021). In conclusion, it is important to explore ways to bring intersectionality into monitoring and the assessment of progress toward the SDGs.

Digital transitions have multiple impacts on women. Better access to technology increases access to information, personal agency, and opportunities for employment and decent work. SDG Goal 8 (sustainable economic growth) looks at digital infrastructure and the relationship between formal and informal employment. A relevant indicator from the perspective of the digital transition is 8.3.1, which measures the "proportion of informal employment in total employment, by sex." Despite such monitoring, we still see gender pay gaps (International Trade Union

Confederation (ITUC) (n.d.)), gender biases against women in the workplace (ILO 2017), and women taking on a disproportionate amount of household chores globally (Soraya and Marta 2021).

Women are under-represented in digital jobs, including top management positions and in academia. For example, men are four times more likely than women to be ICT specialists. Female-owned start-ups receive 23 percent less funding, and are 30 percent less likely to have a positive outcome compared to male-owned businesses (OECD 2018).

The digital transition offers clear opportunities for the economic and social betterment of women, yet the SDG framework does not feature distinct targets (Plan International 2018). Digital platforms/financial services help bridge the gender gap by offering women employment opportunities, the possibilities of extra income, and better access to knowledge and information (Lee and Pollitzer 2016). Financial inclusion is being accelerated by the digital revolution, but there is still a gender gap with regard to account ownership.

Digital transformation provides new avenues for the economic empowerment of women and can contribute to greater gender equality. Digital access helps women claim a larger share of the public space, where they are able to access uncensored information, create social capital via networking, and campaign for social causes (The World in 2050, 2019). Gender equality promotes economic growth and labor productivity and recognizes that properly rewarding the work of women leads to job creation and economic growth (UN Women 2018). Gender equality initiatives will need to foster inclusivity in the labor market to better tap the potentials that women can offer.

Our discussion will now consider indicator 8.3.1.—the only measurement of Goal 8 that is relevant to digital transitions, since it monitors the "proportion of informal employment in total employment, by sex".

Creating jobs in the formal sector is an important strategy to reduce informal employment. It is important to continue to promote and protect the rights of both women and men engaged in the informal sector by improving earnings and working conditions and extending social security. Trends for SDG indicator 8.3.1. across Ethiopia, India, and Australia show that more women than men are employed in the informal sector, with a significant majority from both sexes in informal employment in the agriculture sector (United Nations Department of Economic and Social Affairs 2022).

In Australia, a much higher proportion of informal agricultural workers are women than men (81 percent vs. 68 percent). The proportion of women and men working in non-agricultural sectors is much closer (27 percent vs. 25 percent), as they are across all employment (28 percent vs. 27 percent). In India in 2017–2018, the proportion of women engaged in the agricultural workforce was higher than men (62.2 percent vs. 37.8 percent) (Government of India 2022). However, in Ethiopia, the proportion of women in the agricultural sector is lower (47 percent vs. 53 percent) than men (Europa.eu 2021).

SDG 9 aims to "build resilient infrastructure, promote sustainable industrialization and foster innovation." It has long been recognized that growth in productivity/incomes and improvements in health and education require investment in infrastructure. Inclusive and sustainable industrial development is key to rapidly improving living standards for all people. Despite this, not a single indicator in SDG 9 is gender-disaggregated.

In summary, gender inequality in education and employment affects economic growth by reducing the amount of human capital in a society, which harms economic performance and attenuates the benefits of digital transitions. Closing the gender gap in the formal labor market—through gender-equal access to resources, assets, and technologies—can help improve productivity, create economic diversification, and increase global employment by 189 million (or 5.3 percent) by 2025.

Gender Equality, Technology, and Development

Information and communications technology (ICT) is increasingly important in many aspects of economic and social life, and ensuring that technology develops in a gender equal way is thus of some urgency. This is especially the case for mobile phones and other portable technologies. The Global System for Mobile Communications Association (GSMA), for example, has shown that increasing women's digital/financial inclusion and redressing gender inequalities can increase women's economic opportunities in low- and middle-income countries (LMICs), leading to reductions in poverty and improved quality of life (GSMA 2021).

Research has shown that in LMICs mobile phones can effectively increase access to crucial services, increase agricultural productivity and GDP, empower marginalized groups, and reduce existing inequalities. The GSMA, for example, analyzing data from 18 LMICs estimated that reducing gender disparities in mobile internet use could contribute an extra US$700 billion to their total GDP growth (GSMA 2019a, b). The

GSMA Connected Women Program's *Mobile Gender Gap* Report (2019) showed that both women and men substantially benefit from owning mobile phones. For the majority (60 percent), owning a phone increased safety, improved efficiency in daily activities, and improved access to critical information and value-added services that hitherto had not been available/accessible to them (GSMA 2019a, b).

Similarly, the recent growth in the mobile telephony industry has increased digital inclusion for the previously financially excluded and underserved, especially women and youth in rural areas of LMICs (Bill and Melinda Gates Foundation 2019). The use of a mobile phone (or other digital device/financial service) opens up and increases access to banking facilities for women previously without access (Demirgüç-Kunt et al. 2017). Utilizing evidence from Kenya, Suri and Jack (2016) reported the positive impact of mobile telephony on the livelihoods of the rural poor, especially women, in terms of opening up and securing sustainable economic opportunities. In addition, according to Mittal (2016), when mobile phones were able to deliver context-specific information to male and female farmers in the Indian states of Haryana and Bihar, there was an increased adoption of gender-specific climate-smart agricultural technologies.

Despite its importance, access to—and usage of—mobile technology remains persistently unequal. The GSMA Connected Women Program's more recent *Mobile Gender Gap* report (2020) provides evidence of the disparities between men and women's ownership of mobile phones in LMICs (GSMA 2019a, b). It reports that across LMICs, women are eight percent less likely than men to own mobile phones, and 20 percent less likely to use the internet on mobile devices—the equivalent of about 327 million fewer women than men (OECD 2018).

2.4 Concluding Remarks and Recommendations

Gender and the Green Transition

Since gender-specific targets/indicators in the SDGs are very limited, especially those that relate to the green transition, a new approach is needed.

The UNFCCC recommends:

- integrating gender considerations into adaptation and mitigation strategies;
- fostering a gender-sensitive approach to strengthen institutional, systemic, and individual capacity-building;
- considering gender when developing resource mobilization strategies and applying climate finance instruments; and
- increasing gender mainstreaming in terms of access to—and training on—environmentally sound technologies.

These recommendations could be developed into environmental-themed SDG indicators, with specific targets that could also be reflected in each country's NDCs.

Policymakers need to better serve the needs of different groups. Gender-disaggregated data needs to be gathered more systematically, including on climate change impacts. National and international green and climate change initiatives aim to achieve gender equality, but currently, there is a paucity of gender-disaggregated indicators. It is important for governments to capture such indicators using a gendered and intersectional lens. Disaggregated climate data will help us better understand the different livelihood impacts on men and women, the different economic opportunities in green initiatives, as well as gender-differentiated impacts on health.

Women can play an important role at the local, national, regional, and global levels by advocating for climate justice and a greener environment. Support for women's initiatives needs to be increased, especially where they intersect with initiatives for youth and vulnerable groups. Strategies to progress Goal 5 could be enhanced to support female-led projects that advance climate adaptation and mitigation strategies. All this would benefit from coordinated action by governments, the private sector, and civil society.

Gender-responsive budgeting and financial allocations are also needed to ensure that the way climate change impacts different groups are properly identified and addressed. For instance, target 5.c.1. (which measures the "proportion of countries with systems to track and make public allocations for gender equality and women's empowerment") could be extended to capture budgetary allocations for climate-related programs and initiatives. This can only happen in an enabling environment where local government bodies not only support gender-responsive budgeting but understand their roles and responsibilities with regard to implementing it (Climate and Development Knowledge Network or CDKN 2021).

The Climate Public Expenditure and Institutional Review (CPEIR) methodology helps countries to assess and review their climate-related expenditures (UNDP 2015). Tools like Climate Budget Tagging (CBT) can help to identify, classify, weigh, and mark climate-relevant expenditures in the fiscal budget, thereby enabling monitoring and tracking. Gender and other differential markers (like income levels) could also be tagged using the CBT tool, so that governments can review and recognize the way budgetary support impacts different groups. Gender impact assessments can also help analyze how climate policies interact with gender, as well as identify ways to strengthen and ensure effective gender action and policies (Sauer and Stieß 2021). It will also help to show how core policies and measures can be improved to progress gender justice (Alber n.d.).

Gender and the Digital Transition

The disparities between women's and men's ownership, access to, use of, and benefits from mobile technologies need urgent attention if we are to rapidly achieve gender equality and fully realize the potentials of all people in all countries. We need urgently to narrow the gender gap in mobile ownership and to generate sex-disaggregated data (for mobile technologies and ICT) so that policies, plans, and strategies can fully promote inclusive development. Such data might best be included in routine national censuses. Understanding the relative roles of women and men in terms of urban waste management will also foster gender inclusion in green transitions.

There are other key areas that would benefit from applying a gendered/intersectional lens. For example:

- Analyzing gendered work roles and health risks (including awareness of—and exposure to—chemicals)
- Analyzing gender differences and similarities across countries and industries to make comparisons and learn from other countries' best practices
- Analyzing the gendered impacts of privatizing and commercializing natural resources (such as water, land, and forests)

- Analyzing factors that influence consumptions patterns (including gender, household type, age, lifestyle orientation, socio-demographic conditions, and other intersectional vulnerabilities).

2.5 Implications of the Lack of Sex-Disaggregated Indicators and Statistics

The importance of collecting, analyzing, and using high-quality sex-disaggregated data as well as other gender-related indicators and statistics (especially in developing countries) cannot be overemphasized in terms of progressing the SDGs. The near absence of sex-disaggregated SDG indicators on climate change and ICT is a glaring omission, and a big failure to acknowledge that gender equality is necessary for inclusive economic and social development

Aggregated data masks gender disparities and trivializes the lived experience and realities of the excluded and marginalized, notably women. Lack of gender data ignores the differential vulnerabilities to climate change as well as its specific impacts on women and other groups. Furthermore, aggregated data fails to respect the unique knowledge of women, indigenous groups, and other community groups or to acknowledge what they can offer in terms of designing and implementing climate resilience/adaptation strategies. Women's realities pertaining to mobile phone technology (ownership, use, access, and benefits) are undermined, undocumented, or not considered in statistics or the policies those statistics inform. Most worryingly, the paucity of sex-disaggregated mobile technology data in the case of developing countries (including Ethiopia and India) undermines the opportunity to sensitize policymakers to the need for gender-transformative policies, plans, and strategies to address gender inequalities in ICT and related areas of education, health, and agriculture.

While the SDG framework is a laudable effort, its lack of gender-disaggregated indicators is a significant drawback to capturing and measuring progress on green and digital transitions. A paradigm shift is needed to integrate gender and intersectionality within the SDGs, so that countries are motivated to recognize the different vulnerabilities, needs, and capabilities of different groups, especially women, in order to make equitable progress and ensure no one is left behind.

References

Alber, Gotelind. n.d. GAMMA gender assessment method for mitigation and adaptation. Accessed on 15 December 2021 https://unfccc.int/files/gender_and_climate_change/application/pdf/gendercc-gamma-tool.pdf.

Australian Government. 2018. Indicator 13.b.1. sustainable development goals. 15 December 2021. https://www.sdgdata.gov.au/goals/climate-action/13.b.1.

———. 2021a. *Australia's Nationally Determined Contribution: Communication 2021.* https://www4.unfccc.int/sites/ndcstaging/PublishedDocuments/Australia%20First/Australia%20Nationally%20Determined%20Contribution%20Update%20October%202021%20WEB.pdf.

———. 2021b. *National climate resilience and adaptation strategy 2021–2025: Positioning Australia to better anticipate, manage and adapt to our changing climate.* https://www.awe.gov.au/sites/default/files/documents/national-climate-resilience-and-adaptation-strategy.pdf.

Bill and Melinda Gates Foundation. 2019. *A G7 partnership for women's digital financial inclusion in Africa. Report prepared at the request of the G7 French Presidency.* https://docs.gatesfoundation.org/Documents/WomensDigitalFinancialInclusioninAfrica_English.pdf.

Blackden, C. Mark. 1999. *Gender, growth, and poverty reduction. Africa region findings & good practice info briefs. no. 129.* https://openknowledge.worldbank.org/handle/10986/9873 Licence: CC BY 3.0 IGO.

Climate and Development Knowledge Network (CDKN). 2021. *Supporting climate action through gender-responsive budgeting in Nepal. Policy Brief. Climate and Development Knowledge Network.* https://cdkn.org/wp-content/uploads/2021/03/Mainstreaming-Gender-Responsive-Budgeting_25Mar2021.pdf.

Demirgüç-Kunt, Asli, Leora Klapper, Dorothe Singer, and Saniya Ansar. 2017. *The global findex database 2021: Financial inclusion, digital payments, and resilience in the age of COVID-19.* The World Bank. https://doi.org/10.1596/978-1-4648-1897-4.

Encarnacion, Jessamyn, and Samrat Maskey. 2021. We now have more gender-related SDG data than ever, but is it enough? UN Women Data Hub. 26 May 2021. https://data.unwomen.org/features/we-now-have-more-gender-related-sdg-data-ever-it-enough.

Europa.eu. 2021. Ethiopia Gender Profile Country and Sector March 2021. https://europa.eu/capacity4dev/file/126063/download?token=MVJr8aVp

Federal Democratic Republic of Ethiopia. 2019. *Ethiopia's climate resilient green economy: National adaptation plan.* https://www4.unfccc.int/sites/NAPC/Documents/Parties/NAP-ETH%20FINAL%20VERSION%20%20Mar%202019.pdf.

———. 2021. *Updated nationally determined contribution* https://www4.unfccc.int/sites/ndcstaging/PublishedDocuments/Ethiopia%20First/Ethiopia%27s%20updated%20NDC%20JULY%202021%20Submission_.pdf.

Fletschner, Diana, and Lisa Kenney. 2014. Rural women's access to financial services: Credit, savings, and insurance. In *Gender in Agriculture*, ed. Agnes R. Quisumbing, Ruth Meinzen-Dick, Terri L. Raney, André Croppenstedt, Julia A. Behrman, and Amber Peterman, 187–208. Dordrecht: Springer Netherlands. https://doi.org/10.1007/978-94-017-8616-4_8.

Government of India. 2016. *India's intended nationally determined contribution: Working towards climate justice*. https://www4.unfccc.int/sites/ndcstaging/PublishedDocuments/India%20First/INDIA%20INDC%20TO%20UNFCCC.pdf.

———. 2022. Labour and employment statistics. https://dge.gov.in/dge/sites/default/files/2022-08/Labour_and_Employment_Statistics_2022_2com.pdf

GSMA. 2019a. *Connected women. The mobile gender gap report 2019*. https://www.gsma.com/mobilefordevelopment/wp-content/uploads/2019/03/GSMA-Connected-Women-The-Mobile-Gender-Gap-Report-2019.pdf.

———. 2019b. *The mobile gender gap: Africa*. https://www.gsma.com/mobilefordevelopment/wp-content/uploads/2019/07/The-Mobile-Gender-Gap-in-Africa.pdf

———. 2021. *State of the industry report on mobile money 2021*. https://www.gsma.com/mobilefordevelopment/wp-content/uploads/2021/03/GSMA_State-of-the-Industry-Report-on-Mobile-Money-2021_Full-report.pdf.

ILO. 2017. ACT/EMP research note breaking barriers: Unconscious gender bias in the workplace. https://www.ilo.org/wcmsp5/groups/public/%2D%2Ded_dialogue/%2D%2D-act_emp/documents/publication/wcms_601276.pdf

International Trade Union Confederation (ITUC). n.d.. ITUC economic and social policy brief: The gender wage gap. https://www.ituc-csi.org/IMG/pdf/the_gender_wage_gap_en.pdf

IPCC. 2014. Summary for policymakers. In *Climate change 2014: Impacts, adaptation, and vulnerability. Part A: Global and sectoral aspects. Contribution of working group II to the fifth assessment report of the intergovernmental panel on climate change*, edited by Dokken, David, Christopher Field, Vicente Barros, Katharine Mach, Michael Mastrandrea, M.TE Bilir, Monalisa Chatterjee,K.L. Ebi, Y.O. Estrada, R.C. Genova, B. Girma, E.S. Kissel, A.N. Levy, S. MacCracken, P.R. Mastrandrea, and L.L. White, 1–32. United Kingdom and New York, NY, USA: Cambridge University Press. https://www.ipcc.ch/site/assets/uploads/2018/02/ar5_wgII_spm_en.pdf.

Lee, H., and Pollitzer, E. (eds.). 2016. Goal 8 promote inclusive and sustainable economic growth, employment and decent work for all. In *The role of gender-based innovations for the UN sustainable development goals: Toward 2030: Better Science and Technology for all (edition 1)*, 22–23. Korea Center for Women in

Science, Engineering and Technology (WISET). https://gender-summit.com/images/GS6Docs/SDG_Report_FINAL.Jan13.pdf.

Mittal, Surabhi. 2016. Role of mobile phone-enabled climate information services in gender-inclusive agriculture. *Gender, Technology and Development* 20 (2): 200–217. https://doi.org/10.1177/0971852416639772.

Nabalamba, Alice, Maurice Mubila, and Patricia Alexander. 2019. Economic brief - climate change, gender and development in Africa. *African Development Bank Group*. https://www.afdb.org/en/documents/document/economic-brief-climate-change-gender-and-development-in-africa-25666.

NITI Aayog. 2021. *SDG India index & dashboard 2020–21: Partnerships in the decade of action*. National Institution for Transforming India. Retrieved from https://www.niti.gov.in/writereaddata/files/SDG_3.0_Final_04.03.2021_Web_Spreads.pdf.

OECD. 2018. *Bridging the digital gender divide: Include, upskill, innovate*. Retrieved from Organization for Economic Co-operation and Development. https://www.oecd.org/digital/bridging-the-digital-gender-divide.pdf.

———. 2021. *Gender and the environment: Building evidence and policies to achieve the SDGs. OECD*. https://doi.org/10.1787/3d32ca39-en.

———. n.d.. *Gender, inclusiveness and the SDGs*. Organisation for Economic Co-operation and Development. https://www.oecd.org/environment/gender-inclusiveness-and-sdg.htm.

OHCHR. n.d. Gender-responsive climate action. OHCHR and climate change. *OHCHR*. Accessed 21 December 2021. https://www.ohchr.org/en/climate-change/gender-responsive-climate-action.

Porter, Gina, Kate Hampshire, Albert Abane, Alister Munthali, Elsbeth Robson, Ariane De Lannoy, Augustine Tanle, and Samuel Owusu. 2020. Mobile phones, gender, and female empowerment in sub-Saharan Africa: Studies with African youth. *Information Technology for Development* 26 (1): 180–193. https://doi.org/10.1080/02681102.2019.1622500.

Price, Roz. 2018. *Women-initiated measures to cope with environmental stresses and climate change in South Asia. Helpdesk report*. Institute of Development Studies. https://assets.publishing.service.gov.uk/media/5aafce6b40f0b64ab1a1e876/Women-initiated_measures_to_cope_with_environmental_stresses_in_South_Asia.pdf.

Roser, Ritchie, and Mispy Ortiz-Ospina. 2018. Measuring progress towards the sustainable development goals: Goal 5: Gender equality - SDG tracker. *Our World in Data*. 2018. https://sdg-tracker.org/gender-equality.

Sami, Neha, Chandni Singh, and Amir Bashir Bazaz. 2016. *Climate Change Policy in India and Goal 13*. Indian Institute for Human Settlements. https://doi.org/10.24943/updsdg2016_1.

Sauer, Arn, and Immanuel Stieß. 2021. Accounting for gender in climate policy advice: adapting a gender impact assessment tool to issues of climate change.

Impact Assessment and Project Appraisal 39 (3): 262–273. https://doi.org/10.1080/14615517.2021.1904710.

Seck, Papa, and Samrat Maskey. 2019. Use data to take the guesswork out of policymaking. UN Women Data Hub. 24 September 2019. https://data.unwomen.org/features/use-data-take-guesswork-out-policymaking.

Sender, John, and Sheila Smith. 2012. *Poverty, class and gender in rural Africa*. 1st ed. Routledge. https://doi.org/10.4324/9780203835487.

Soraya, Seedat, and Rondon Marta. 2021. Women's wellbeing and the burden of unpaid work. *BMJ* 374: n1972. https://doi.org/10.1136/bmj.n1972.

Suri, Tavneet, and William Jack. 2016. The long-run poverty and gender impacts of mobile money. *Science* 354 (6317): 1288–1292. https://doi.org/10.1126/science.aah5309.

Terry, Geraldine. 2009. No climate justice without gender justice: An overview of the issues. *Gender and Development* 17 (1): 5–18.

UN Women. 2016. *Leveraging co-benefits between gender equality and climate action for sustainable development mainstreaming gender considerations in climate change projects*. https://unfccc.int/files/gender_and_climate_change/application/pdf/leveraging_cobenefits.pdf

———. 2018. *Gender equality as an accelerator for achieving the sustainable development goals*. Discussion Paper. www.undp.org/publications/gender-equality-accelerator-achieving-sdgs.

———. 2019. *Progress on the sustainable development goals: The gender snapshot 2019*. https://www.unwomen.org/sites/default/files/Headquarters/Attachments/Sections/Library/Publications/2019/Progress-on-the-SDGs-The-gender-snapshot-2019-two-page-spreads-en.pdf.

———. 2021. *Progress on the sustainable development goals: The gender snapshot 2021*. https://unstats.un.org/sdgs/gender-snapshot/2021/UNW_GenderSnapshot.pdf.

———. n.d.-a. Australia: Country fact sheet | UN Women Data Hub. Accessed 16 December 2021. https://data.unwomen.org/country/australia.

———. n.d.-b. Australia: Country fact sheet | UN Women Data Hub. Accessed 16 December 2021. https://data.unwomen.org/country/australia.

———. n.d.-c. Ethiopia: Country fact sheet | UN Women Data Hub. Accessed 16 December 2021. https://data.unwomen.org/country/ethiopia.

———. n.d.-d. India: Country fact sheet | UN Women Data Hub. Accessed 16 December 2021. https://data.unwomen.org/country/india.

———. n.d.-e Women count: Data. UN Women Data Hub. Accessed 20 December 2021. https://data.unwomen.org/data-portal.

UNDP. 2015. A methodological guidebook: Climate Public Expenditure and Institutional Review (CPEIR) | United Nations Development Programme. UNDP. 2015. https://www.undp.org/asia-pacific/publications/methodological-guidebook-climate-public-expenditure-and-institutional-review-cpeir

UNICEF and UN Women. 2013. *Global thematic consultation on the post-2015 development Agenda: Addressing inequalities synthesis report of global public consultation.* https://genderandenvironment.org/wp-content/uploads/dropbox/Publication//Addressing%20Inequalities%20Synthesis%20Report%20of%20Global%20Public%20Consultation.pdf

United Nations Department of Economic and Social Affairs. 2022. 'Global SDG indicators data platform'. 2022. https://unstats.un.org/sdgs/dataportal.

United Nations Statistics Division (UNSD). 2021. *SDG indicator metadata: Harmonized metadata template* (format version 1.0.). https://unstats.un.org/sdgs/metadata/files/Metadata-13-0b-01.pdf.

USAID. 2021. *A briefer: Gender inequality causes poverty.* https://banyanglobal.com/wp-content/uploads/2021/03/Gender-Inequality-Causes-Poverty-Briefer.pdf

Women and Gender Constituency. 2017. Resolve trash to cash: Providing social and economic security to ostracised women. https://womengenderclimate.org/gjc_solutions/resolve-trash-to-cash-providing-social-and-economic-security-to-ostracised-women/.

WomenWatch. 2009. *Women, gender equality and climate change. Fact sheet.* https://www.un.org/womenwatch/feature/climate_change/downloads/Women_and_Climate_Change_Factsheet.pdf

World Bank. 2012. *Climate change budget tagging: A review of international experience.* https://openknowledge.worldbank.org/bitstream/handle/10986/35174/Main-Report.pdf?sequence=1&isAllowed=y

CHAPTER 3

Gender Inequalities, Poverty, and Disparities: Impact and Links in the Era of Digital Transformations and Green Transition

Amira Kaddour and Hela Ghbara

3.1 Introduction

Eradicating poverty, in all its forms, is key goal of development. Poverty reduction requires growth strategies and distributive policies, the combination of which is specific to each country. The Poverty-Growth-Inequality triangle states that any development strategy is totally dependent on the rate of growth and variations in distribution within the population. Since basic needs are subject to changes over time and space, the absolute poverty line may vary from country to country and may also vary within a

A. Kaddour (✉)
National School of Advances Sciences and Technologies, University of Carthage, Amilcar, Tunisia

Includovate, Tunis, Tunisia
e-mail: amira.kaddour@enstab.ucar.tn

H. Ghbara
High School of Business, University of Manouba, Manouba, Tunisia

K. Drucza et al. (eds.), *Centering Gender in the Era of Digital and Green Transition*,
https://doi.org/10.1007/978-3-031-38211-6_3

single country over long periods of time. Digital transformations can drive great services yet increase exclusion if the specific needs of vulnerable people are not considered. It is thus crucial to take gender aspects into account in the fight against poverty. Gender equality has long been an issue that needs additional efforts in many countries. Digital transformations have given rise to new technologies, services, and opportunities that have made it easier to access and use resources, but disparities in access remain. Despite the potential of technology, poverty and gender inequality remain major problems in many countries, with women and children bearing the brunt of their burdens. While digital transformations have offered new ways to tackle poverty and gender inequality, it has also created new challenges (and highlighted existing ones). To truly make progress, governments need to invest in programs that focus on both economic and social empowerment of women and girls, while also tackling the root causes of poverty and inequality. This includes investing in education, healthcare, and economic opportunities to ensure that everyone has access to the same resources and opportunities. It is only then that we can truly see real progress in gender equality and poverty reduction.

3.2 The Poverty-Growth-Inequality Triangle: Considering Equality Toward Inclusion and Long-Term Growth

According to a recent study by the International Monetary Fund (IMF 2021), the relationship between growth, poverty, and inequality is complex and multifaceted. While economic growth has the potential to effectively reduce poverty, its impact on inequality is uncertain and varies depending on the diverse factors which drive growth. The effects of poverty and inequality on economic growth are also ambiguous and influenced by multiple channels. However, evidence suggests that poverty and inequality can hinder long-term growth prospects.

The IMF study underscores the importance of policy interventions in shaping these interdependencies. Policies aimed at enhancing equality of opportunity can simultaneously promote inclusivity and foster long-term economic growth. Researchers have reached a broad consensus on the need to address equity and income distribution issues alongside economic growth. Indeed, Fambon (2005) highlights the significance of considering the most disadvantaged groups of the population in relation to economic growth. The Fambon 2005 study emphasizes that economic strategies must align with poverty reduction strategies, and highlights the potentially harmful effects of adjustment programs on certain social strata. To pragmatically assess poverty alleviation efforts, therefore, measuring growth's impact on poverty and social inequalities is crucial.

The relationship between growth, inequality, and poverty has been a recurring theme in economics. While the debate has traditionally focused on the relationship between growth and inequality, the inclusion of poverty reduction strategies has expanded the discussion. Several studies, including Deininger and Squire (1996), Dollar and Kraay (2002), and Ravallion (2002), highlight the importance of focusing on the relationship between growth and distribution within this triangle. Understanding this relationship is essential as it affects other dynamics and acts as a prerequisite for successful poverty reduction.

Fambon (2005) proposes that poverty levels depend on average income and income inequality. Consequently, the decline in absolute poverty is influenced by two factors: an increase in average income, assuming a constant relative income distribution, and any redistribution of income toward the poor at a constant average income. Therefore, changes in poverty can be fragmented into two effects: a growth component related to changes in average income, and an inequality component due to changes in income distribution. Analyzing the levels of these two components provides insights into the relative impacts of poverty reduction through growth and improved equality.

3.3 Place of Digital Transformations

Digital transformations give developing countries a tremendous opportunity to foster growth and become part of the globalized economy. According to Sorgner and Cabot (2019), in recent years, digital transformations have become an increasingly essential component and a powerful catalyst for economic development. They consist of all the techniques and devices put in place to transmit, exchange, store, and process information. Basically, it is a sophisticated system that is used to manipulate information into data that is easy to manage and communicate. Digital transformation also uses microelectronics, telecommunications, and above all the world of the Internet, which can be considered as a sector, with online commerce and all forms of virtual communication. Sorgner and Cabot (2019) indicate that digital transitions can transform global communication networks and open up excellent prospects for African countries in terms of poverty reduction. They play an indirect role in promoting economic growth and can also have a remarkable impact on standards of living, enabling the poor to access markets, demand services, receive an education, and learn new skills.

Over the past ten years, digital transformations have changed the way we live, how we acquire knowledge, and how we produce. Kouadio and Gakpa (2022) report that technological advances have raised hopes, especially in developing countries, that these tools could accelerate their economic development. Indeed, thanks to digital transformations, cooperation networks between international researchers have been strengthened, and international companies have seen the possibilities of their access to the international market increase, and to digitize delivered services. The improvement of human capacities or of social, economic, political, and cultural conditions is also considered in terms of what digital transformation can bring to us. Information and communication technologies benefit from a quasi-consensus on the positive role they play in individual and collective development. This is not surprising since the evolution of technology has often been considered an important engine of the evolution of the economy and society. Among the most famous reflections on the role of technology are those of Karl Marx, who considers that technical evolution forms the “infrastructure” of society. From this infrastructure flow relations of production and social relations in general, as well as associated value systems, which form the “superstructure” of society. The impact of these new digital technologies is profound and wide-ranging. Asongu et al. (2021a) discuss how email allows students to share knowledge and collaborate with other students around the world. The effective use of information technology in education in countries obviously involves the appropriation of computers, office automation tools, or more specific learning. For some time now, a new concept of technology in education has been developing—E-Learning—which represents a distance learning process based on multimedia resources and allows one or more people to learn directly from their computers. Asongu and Al (2021b) report that digital transformations are improving the quality of health services, playing a critical role in the administration and management of information, as well as the dissemination of medical discoveries. New digital technologies also make it possible to monitor the environment through an almost unlimited amount of data. Geographic information systems, which facilitate the collection, storage, and analysis of information, are an effective means in the development of viable management plans and support the prediction of environmental threats.

Agriculture is one of the most diversified economic sectors since it includes farmers, farmers' organizations, government agencies, research institutes, traders, multinationals, etc. (Deichmann et al. 2016). The sector depends on a fruitful and equitable interaction between its many different actors, and digital technologies can facilitate this by easing communication, and exchange of information, and help improve access to information on prices, markets, and productions.

A World Bank report (Finance for equitable recovery) from 2022 confirms how digital technologies can foster inclusion, efficiency, and innovation. More than 40% of adults in East Africa pay their utility bills by mobile phone, for instance. In China, 8 million entrepreneurs—a third of whom are women—use e-commerce platforms to sell products domestically and export them to 120 other countries. India has provided a unique digital identity to almost a billion people in five years and has widened access to public services and reduced corruption by doing so. Nevertheless, according to Kiyindou (2010), information and communication technologies, even if they can help reduce poverty, are only tools and cannot—on their own—solve complex problems like poverty. Digital transformations can thus be one instrument for eradicating poverty but not as a substitute or replacement tool for other development measures. Indeed, poverty represents the lack of resources necessary to maintain a minimum standard of living. It results from unequal access to political, social, and economic resources. The involvement of digital transformations in development projects results in a change within society. The latter is based on information and knowledge, which are essential to the development process. However, not having access to these sources creates a gap that undermines the prospects for economic growth, development as well as distribution of wealth, and social emancipation.

Digital transformations can accelerate development, contribute to economic growth, bring people together, and enable better use of resources. Yet many countries are struggling to invest in innovation and digital technologies because they cannot afford them. Nearly 2.9 billion people around the world, especially women and girls, do not have a phone, computer, or internet access (UN 2021). According to Opp and Lagunas (2022), digital technologies are a powerful driver of gender equality, enabling women and girls to access new information, opportunities, and resources. However, the gender digital divide persists, in part because of deeply entrenched social and gender norms and gender-specific stereotypes. As a result, many women—particularly those in developing countries—continue to face challenges in accessing and using digital technologies. The rate at which girls and young women learn to code, for example, remains lower than that of boys and men. Opp and Lagunas (2022) show that globally, women make up only 29% of the Science, Technology, Engineering, and Mathematics (STEM) workforce and are less likely to be employed than men.

3.4 Impact of the Digital Gap on Economic Inclusion and Poverty

The gender gap in technology refers to the differences between men's and women's access, adoption, and use of technology. This gap can manifest in various ways, such as a lack of access to technology for women, a lack of digital literacy or technological skills among women, or a lack of representation of women in technology-related fields. This gender gap can have a significant impact on women's economic and social opportunities, as well as on their ability to fully participate in the economy. It can also exacerbate existing gender inequalities. A variety of potential factors, including social norms, contribute to the digital gender gap. For generations, social norms have made men responsible for the technological aspects of daily life, relegating many women to non-technological domestic roles. Social norms that influence whether a woman receives a higher education or can keep a job outside the home also impact the use of digital technology. Thus, though digital technologies can transform our economies, societies, and public institutions, these changes are neither given nor automatic. Breunig and Majeed (2020) believe that countries that invest in both digital technologies and their analogue complements will reap large dividends, while others will likely remain on the side-lines. Without a solid foundation, technologies risk creating economic disparities, increasing inequalities, and opening the door to government interference.

According to Deichmann et al. (2016), wealthy, skilled, and influential people around the world who have benefited from rapid digital expansion are better placed to take advantage of new technologies. There are many individual success stories, but so far, the impact of technologies on overall productivity, increased opportunities for the poor and middle class, and the spread of responsible governance has not lived up to expectations.

Digital technologies are spreading rapidly, but their dividends—growth, jobs, and services—are slow to follow. Deichmann et al. (2016) show how the digital revolution is transforming the world, facilitating information flows and the rise of developing countries that are able to take advantage of these new opportunities. The fact that today 40% of the world's population is connected through the internet is a staggering transformation. While this achievement is to be welcomed, it is also an opportunity to remember that we must avoid creating a new underclass. In a world where almost 20% of the global population is unable to read and write, the diffusion of digital technologies alone is unlikely to close the knowledge gap. Poverty thus becomes a deprivation of basic capacities and not just a simple low income. According to Kiyindou (2010), a significant part of poverty is linked to diseases or population movements. This is where the reframing comes in. This should take place not only in general policies but also in the content delivered by technological devices. Therefore, in the field of information and communication technologies, the reduction of what is commonly called the "digital divide" becomes a question of social justice. The improvement of living conditions through information and communication technologies depends first on the resolution of several internal, demographic, institutional, and productive challenges. To fulfill the promise of developing a new digital era, the World Bank proposes two main measures: firstly, to reduce the digital divide by making the internet universal, affordable, open, and secure; and second, to strengthen regulations that ensure business competition, adapt workers' skills to the demands of the new economy, and promote accountable institutions.

Erlando et al. (2020) assert that to benefit the most from digital transformations, countries must create an enabling environment for technologies. That includes fostering regulations that facilitate competition and market entry, skills that allow workers to take advantage of the digital economy, and institutions that are accountable to citizens. Digital technologies can, in turn, accelerate development.

3.5 The Economic Benefits of Diversity in Digital and Green Transition

Diversity in the workplace can lead to several economic benefits. One of the main benefits is that it can lead to increased innovation and creativity, as a diverse group of people will bring different perspectives and ideas to the table. This can result in improved decision-making and problem-solving, which can lead to increased productivity and profitability. In addition, studies have shown that companies with more diverse workforces tend to have better financial performance than those without. For example, a study by McKinsey and Company found that companies in the top quartile for ethnic and racial diversity were 35% more likely to have financial returns above their respective national industry medians. Moreover, diversity can also help companies better serve diverse customer bases, as a diverse workforce will better understand the needs and preferences of different groups of people.

Studies have also shown that having more women in leadership positions can lead to better performance for companies. For example, a study by Peterson Institute for International Economics found that companies with more women in top leadership positions had a 34% higher return on equity than those with fewer women in such positions. In addition, research has consistently shown that companies with more gender-diverse management teams are more likely to outperform their peers in terms of financial performance.

3.6 The Sustainable Development Goals with a Gender Lens

The main objective of the Sustainable Development Goals (SDGs) is to address and overcome key global issues such as poverty and inequality. However, to fully address and progress these issues, it is critical to consider the intersectionality of the green transition and digital transformations with a gender lens. This means understanding and addressing how these issues disproportionately impact women and marginalized groups and incorporating a gender-responsive approach into efforts to achieve the SDGs. This can include, for example, ensuring equal access to green technologies and digital resources for men and women, and addressing the specific needs and challenges faced by women in the transitions to a sustainable future.

In 2022, UN Women released its report, *Progress on the Sustainable Development Goals: the Gender Snapshot* (2022). This looks through a gender lens to summarize our current global situation, highlighting progress and challenges to achieving the SDGs. Table 3.1. analyzes in detail the current situation in relation to each specific goal, and then provides their impact.

Table 3.1 Looking at the SDGs Through a gender lens (Situation Source: *Progress on the Sustainable Development Goals: the Gender Snapshot* (UN Women 2022), impact provided by authors)

Goal	*Statistics*
1	**Situation:** Globally, over 380 million women and girls are in extreme poverty, living on less than $1.90 a day. If current trends continue, in sub-Saharan Africa more women and girls will live in extreme poverty by 2030 than do today. **Impact:** Poverty has significant negative impacts on women's lives and well-being, limiting their access to basic needs such as food, water, and healthcare, as well as limited opportunities for education and employment. Extreme poverty also increases the risk of violence, exploitation, and other forms of oppression. Poverty trends in sub-Saharan Africa are particularly concerning because the region already has one of the highest rates of extreme poverty in the world, and an increase would have devastating consequences for the health, education, and overall well-being of women and girls in the region. It would also impede the region's overall development. (Evidence shows that when women and girls are empowered, their families, communities, and countries benefit.) To address this issue, it is important to implement policies and programs that target the specific needs and challenges faced by women and girls living in extreme poverty, such as increasing access to education, healthcare, and economic opportunities, and addressing the structural inequalities that perpetuate poverty and marginalization.
2	**Situation:** Globally, nearly one in three women experienced moderate or severe food insecurity in 2021. Rising food prices are likely to exacerbate hunger around the world. **Impact:** The impact of food insecurity on women is severe and wide-reaching. Food insecurity refers to a lack of access to enough food for an active and healthy life. When women are food insecure, it affects not only their health and well-being, but also the health and well-being of their families and communities. Food insecurity can lead to malnutrition, which can cause a variety of health problems, including stunted growth in children, anemia, and an increased risk of infectious diseases. It can also lead to poor cognitive development and reduced educational attainment. Additionally, women who are food insecure may have to make difficult choices—such as reducing the quantity and quality of food they consume—and may have to spend more time and energy securing food, which can limit their ability to engage in other productive activities such as working or caring for their families. Recent rising food prices are a major concern as they are likely to exacerbate food insecurity around the world, particularly for those living in poverty. High prices can make it difficult for low-income families to afford enough nutritious food and can also increase their risk of hunger. To address this issue, it is important to implement policies and programs that increase access to nutritious food, such as food assistance programs and income support, and address the underlying structural issues that contribute to food insecurity, such as poverty and lack of economic opportunities. Additionally, investing in sustainable agriculture and promoting food security for smallholder farmers, particularly women and girls is crucial, as it can increase the availability, affordability, and accessibility of nutritious food.

(*continued*)

Table 3.1 (continued)

Goal	*Statistics*
3	**Situation:** Unsafe abortion is a leading but preventable cause of maternal mortality and morbidity. Today, over 1.2 billion women and girls of reproductive age live in countries and areas with some restrictions on access to safe abortion. About 102 million live in places where abortion is prohibited altogether. **Impact:** Unsafe abortion is a significant public health concern, and a leading cause of preventable maternal mortality and morbidity. Unsafe abortions are those that are performed by unskilled individuals, or in an environment that does not meet minimum medical standards, and can lead to serious health complications, including infection, bleeding, and death. Restrictions on access to safe abortion can have serious consequences for women and girls, as they may be forced to seek unsafe, illegal abortions, which can lead to serious health complications and even death. Additionally, in places where abortion is restricted or prohibited, the stigma and criminalization of abortion can prevent women and girls from seeking the care and support they need. To address this issue, it is important to increase access to safe and legal abortion services, as well as comprehensive sex education and family planning services. This can help to reduce the number of unintended pregnancies and reduce the need for unsafe abortions. Additionally, it is important to work to reduce the stigma and criminalization of abortion, to ensure women and girls have access to the care and support they need, including post-abortion care if needed.
4	**Situation:** Protracted conflicts and wars keep girls out of school, and perpetuate gender gaps in terms of access to school and learning. 54% of girls who are not in formal education worldwide live in crisis-affected countries. **Impact:** Protracted conflicts, wars, and other forms of violence can have a significant impact on girls' access to education and learning. When conflicts occur, schools and other educational facilities can be destroyed, and girls may be forced to flee their homes. Additionally, girls in conflict-affected areas may be at increased risk of violence, exploitation, and other forms of oppression, which can make it difficult or impossible for them to attend school. In some cultures and societies, girls are not valued as much as boys and, as a result, cultural and social norms may perpetuate gender gaps in terms of access to school and learning. This can be particularly true in crisis-affected areas, where resources may be scarce, and girls' education may be seen as less of a priority. Girls' lack of inclusion has a significant impact. Evidence shows that when girls are educated, their families, communities, and countries benefit. Girls' education is a powerful tool for breaking the cycle of poverty and promoting economic growth. Additionally, girls who are not in school are more likely to be married and have children at an early age, which can have detrimental effects on their health and well-being. To address these issues, it is important to invest in educational programs that target the specific needs and challenges faced by girls in crisis-affected areas, such as providing safe learning spaces and increasing access to education for displaced girls. Raising awareness of the importance of girls' education and changing societal norms that discriminate against girls are also crucial.

(*continued*)

Table 3.1 (continued)

Goal	*Statistics*
5	**Situation:** • At the current rate of progress, it may take another 286 years to remove discriminatory laws and close prevailing gaps in legal protections for women and girls. • At the current pace of change, parity will not be achieved for another 140 years. • Globally, more than one in ten women and girls (aged 15–49) are subjected to sexual and/ or physical violence by an intimate partner. • School and daycare closures in 2020 led to an estimated 512 billion additional hours of unpaid childcare globally for women. • The glass ceiling remains intact: Close to one in three managers/supervisors is a woman, and women only hold just over one-third of seats in local decision-making bodies. • In 2021, 4475 communities made public declarations committing to eliminating female genital mutilation. • Only 26% of countries have a comprehensive system to track gender budget allocations. **Impact:** Unpaid care work (housekeeping, childcare etc.) is not recognized as work, not remunerated, not counted in GDP, and not considered when policies are made. As a result, women have less time and resources for other activities, such as working, studying, or participating in civic life. When women are forced to take on additional unpaid care work, it can limit their ability to participate in the paid workforce, which can have long-term consequences for their economic security and opportunities for career advancement. The COVID-19 pandemic has shown how the glass ceiling remains intact. Closures and other disruptions disproportionately affected women's participation in the workforce, set back the progress made in gender equality in the workplace, and resulted in a widening of the gender pay gap. To address this issue, it is important to recognize and value unpaid care work, and to invest in policies and programs that support women to balance unpaid care work with paid work, such as affordable childcare/eldercare, flexible working arrangements, and parental leave policies. Additionally, it is also important to promote gender-responsive policies that consider the specific needs and challenges faced by women in the workforce and to facilitate changes in societal norms and stereotypes that discriminate against women in the workplace.

(continued)

Table 3.1 (continued)

Goal	*Statistics*
6	**Situation:** Lack of clean water claims the lives of more than 800,000 women and girls every year. **Impact:** Lack of clean water is a major public health concern, which disproportionately affects women and girls, and can lead to a variety of health problems, including waterborne diseases, poor hygiene, and poor sanitation. It also increases the risk of violence and exploitation, as women and girls may have to travel long distances to access water and may be at increased risk of harassment and assault. Women and girls in developing countries spend on average six hours a day collecting water, time that could be used more productively for other activities such as working, studying, or participating in civic life. This has a significant impact on their health and well-being, as well as on their economic security and opportunities for career advancement. To address this issue, it is important to invest in infrastructure and programs that increase access to clean water, such as water wells, pumps, and filtration systems. This will help to ensure that the specific needs and challenges faced by women and girls are considered and that their access to clean water is improved.
7	**Situation:** Affordable, clean energy is key to productivity and lifesaving care, but remains out of reach for millions of women and girls in Asia and sub-Saharan Africa. Rising energy prices are making matters worse. **Impact:** Without access to energy, women and girls are often responsible for collecting firewood, which can be time-consuming and dangerous. They may not have access to electricity for powering lifesaving medical equipment or refrigerating vaccines. Additionally, the high cost of energy can strain already limited household budgets, making it difficult for families to afford necessities. Rising energy prices can exacerbate these challenges, making it even harder for low-income families to access energy and improve their quality of life.
8	**Situation:** Women's participation in the labor force in 2022 is projected to remain below pre-pandemic levels in 169 countries and areas. **Impact:** Women's low participation in the labor force can lead to an overall decrease in economic growth and productivity, as well as a loss of human capital and potential income for individuals and families. It can also contribute to income inequality and gender pay gaps. Additionally, women's low participation in the labor force can have a detrimental effect on their social and economic empowerment and limit their ability to make independent decisions, participate in political processes, and have a voice in their communities. The pandemic has significantly reduced women's participation in the workforce, since women are disproportionately represented in sectors and jobs that have been affected by the pandemic, and also due to the pressure of additional care work and home-schooling.

(*continued*)

Table 3.1 (continued)

Goal	*Statistics*
9	**Situation:** Women hold only two in every ten jobs globally in the fields of *science*, technology, engineering and mathematics (STEM). Only 16.5% of inventors associated with a patent are women. **Impact:** Women's limited global representation in STEM limits the diversity of perspectives, ideas, and solutions in those fields, which in turn can impede innovation and progress. It also limits the number of STEM role models for young women and girls and can discourage them from pursuing careers in the field. The gender gap in STEM can also perpetuate the overall gender pay gap and limit women's earning potential and financial independence. The low number of female inventors is also significant, and likely due to a combination of barriers, such as lack of access to education and training, unconscious bias, and discrimination in the workplace. Women's limited participation in STEM means that the sector misses out on diverse perspectives, ideas, solutions, innovations, and creativity. This in turn has an impact on the products and services developed, which may be less inclusive of the needs and perspectives of different groups.
10	**Situation:** By the end of 2021, some 44 million women and girls had been forced to flee their homes due to climate change, war, conflict, and human rights violations. **Impact:** This has far-reaching global implications and can have a devastating effect on the lives of those affected. When women and girls are forced to flee their homes, they are often at greater risk of violence, exploitation, and abuse. They may also be separated from their families, which can cause emotional trauma and long-term mental health issues. Additionally, when women and girls are displaced, they often lose access to basic services such as education and healthcare, which can have a negative impact on their physical and mental well-being. They may also be denied their rights to property and land, which can make it difficult for them to rebuild their lives after displacement. Climate change, war, conflict, and human rights violations are all drivers of displacement. These situations can be particularly harmful for women and girls, as they may be disproportionately affected by the destruction of their homes, communities, and livelihoods. Moreover, displacement can also expose women and girls to increased vulnerabilities, such as gender-based violence, exploitation, and lack of access to basic services.
11	**Situation:** The majority of the world's women live in cities. 49% of women in urban areas report feeling less safe walking alone at night since COVID-19 began. **Impact**: These issues limit women's freedom of movement and access to opportunities. When women do not feel safe in public spaces, they are less likely to participate in activities outside of their homes, such as work, education, and social events. This can decrease their participation in the workforce, and limit their ability to access education and healthcare, among other services. Lack of safety can also have a negative impact on mental health and well-being. Fear of violence and harassment can cause women to experience anxiety and depression, which can have long-term effects on their physical and mental health. Moreover, COVID-19 has made the situation worse, as lockdowns and social distancing measures have left many women isolated and at risk of domestic violence. Additionally, the economic crisis that has ensued from the pandemic has also led to a rise in poverty and unemployment, which can increase the vulnerability of women to violence and exploitation.

(*continued*)

Table 3.1 (continued)

Goal	*Statistics*
12/13/ 14/15	**Situation:** On average, one disaster was recorded daily between 1970–2019, each causing an average of 115 deaths and losses of $202 million. **Impact:** Women, especially from poor and marginalized communities, are disproportionately affected.
16	**Situation:** Strong institutions that support just and peaceful societies are not possible without women. Yet globally women hold just 42% of judicial positions and make up a small proportion of police forces—a mere 16%. **Impact:** Women are victims of gender-based violence and harassment, and supporting the contribution of women as peacebuilder is crucial.
17	**Situation:** Just 4.6% of bilateral allocable ODA goes to programs where gender equality is the main objective. **Impact:** Funding for gender equality is not keeping pace with the increasing severity of global challenges and backlash against women's rights.

3.7 Concluding Remarks and Recommendations

Tackling the interconnected issues of poverty, inequality, and economic growth in the face of digital transformations and green transition challenges requires a comprehensive approach that addresses the structural causes of these problems. One key strategy is to invest in education and skills training to ensure that individuals, particularly those from disadvantaged backgrounds, have the necessary skills to participate in the digital and green economies. Additionally, promoting inclusive economic growth—through policies such as progressive taxation, social safety nets, and targeted investment for disadvantaged communities—is crucial to ensuring that all individuals can participate in and benefit from economic growth. Supporting small and medium-sized enterprises (SMEs) is also important for creating jobs, increasing economic growth, and reducing poverty. Additionally, sustainable and green investments can create new jobs, improve public health, and reduce poverty. Furthermore, addressing systemic issues such as discrimination, lack of access to education and healthcare, and lack of political representation is essential through policies like affirmative action and anti-discrimination laws. Finally, building resilience to shocks—such as pandemics, economic crises, or climate change—through policies that promote social protection, health systems, and disaster risk management is also important for sustainable and inclusive development.

References

Alain Kiyindou. 2010. La lutte contre la pauvreté et/ou pratique des TIC. *Communication & languages*, (N° 163), 121–133. https://www.cairn.info/revue-communication-et-langages1-2010-1-page-121.htm

Alina Sorgner, and John Cabot. 2019. The impact of new digital technologies on gender equality in developing countries. Department of policy, research and statistics. https://www.unido.org/api/opentext/documents/download/16760725/unido-file-16760725

Angga Erlando, Feri Dwi Riyanto, Someya Masakazu. 2020, October. Financial inclusion, economic growth, and poverty alleviation: evidence from eastern Indonesia. *Helyon* 6(10), e05235. https://pubmed.ncbi.nlm.nih.gov/33088971/

Deininger, Klaus, and Lyn Squire. 1996. A new data set measuring income inequality. *World Bank Economic Review* 10 (3): 565–591. https://www.jstor.org/stable/3990058.

Dollar, David, and Aart Kraay. 2002. Growth is good for the poor. *Journal of Economic Growth* 7 (3): 195–225. https://www.jstor.org/stable/40216063.

Fambon, Samuel. 2005. Croissance économique, pauvreté et inégalité des revenus au Cameroun. *Revue d'économie du développement* 13 (1): 91–122. https://www.cairn.info/revue-d-economie-du-developpement-2005-1-page-91.htm.

IMF. 2021. Poverty Reduction and Growth Trust–2020-21 Borrowing Agreements with The Government of Canada as Represented by The Minister of Finance, and The People's Bank of China. https://www.imf.org/en/Publications/Policy-Papers/Issues/2021/07/29/Poverty-Reduction-and-Growth-Trust2020-21-Borrowing-Agreements-with-The-Government-of-463063

Kouadio, Hugues Kouassi, and Lewis-Landry Gakpa. 2022, January–February. Do economic growth and institutional quality reduce poverty and inequality in West Africa? *Journal of Policy Modeling*, 44(1), 41–63. https://www.sciencedirect.com/science/article/abs/pii/S0161893821001095

Ravallion, Martin, and Gaurav Datt. 2002. Why has economic growth been more pro-poor in some states of India than others?. *Journal of Development Economics* 68 (2): 381–400

Robert Breunig, Omer Majeed. 2020 May. Inequality, poverty and economic growth. *International Economics*, 161, 83–99. https://econpapers.repec.org/RePEc:eee:inteco:v:161:y:2020:i:c:p:83-99

Robert Opp et Raquel Lagunas. 2022. Dépasser les stéréotypes, vers l'égalité: transformation numérique et genre. https://www.undp.org/fr/blog/depasser-les-stereotypes-vers-legalite-transformation-numerique-et-genre.

Simplice Asongu, Mouna Amari, Anis Jarboui, and Khaireddine Mouakhar. 2021a. ICT dynamics for gender-inclusive intermediary education: Minimum poverty and inequality thresholds in developing countries. *Telecommunications Policy*, 45(5), 102–125. https://econpapers.repec.org/RePEc:eee:telpol:v:45:y:2021:i:5:s0308596121000306

Simplice Asongu, Yann Nounamo, Henri Njangang, and Sosson Tadadjeu. 2021b. "Gender inclusiveintermediary education, financial stability and female employment in the industry in SubSaharan Africa". *Finance Research Letters* 43(November): 101968 (16).

UN Women. 2022. Poverty deepens for women and girls, according to latest projections. https://data.unwomen.org/features/poverty-deepens-women-and-girls-according-latest-projections

Uwe Deichmann, Aparajita Goyal and Deepak Mishra. 2016, November. Will digital technologies transform agriculture in developing countries? *Agricultural Economics, International Association of Agricultural Economists*, 47(S1), 21–33. https://ideas.repec.org/a/bla/agecon/v47y2016is1p21-33.html

CHAPTER 4

Considering Climate Change Mitigation and Adaptation Through a Gendered Lens

Amira Kaddour, Maissa Louhichi, and Emna Fourati

4.1 Introduction

Although traditional economic models have been successful in generating wealth and promoting economic growth in many countries, significant issues such as poverty, inequality, marginalization, and technological disparities have continued to exist in many parts of the world. The escalating challenges posed by climate change, as well as inequities highlighted by

A. Kaddour (✉)
National School of Advances Sciences and Technologies, University of Carthage, Amilcar, Tunisia

Includovate, Tunis, Tunisia
e-mail: amira.kaddour@enstab.ucar.tn

M. Louhichi
Sfax Faculty of Sciences, Mourou, Tunisia

E. Fourati
Tunis, Tunisia
e-mail: emna.fourati@fss.u-sfax.tn

K. Drucza et al. (eds.), *Centering Gender in the Era of Digital and Green Transition*,
https://doi.org/10.1007/978-3-031-38211-6_4

COVID-19, have prompted a re-evaluation of current economic models. International organizations are, therefore, increasing their efforts to promote more inclusive societies, and more resilient development, as well as decarbonization/mitigation strategies and policies that foster a just transition. An isolationist approach cannot solve the problem of climate change. But an integrationist perspective can do this by addressing the ethical issues and disparities that underpin global injustices, including those related to gender, culture, poverty, and the law.

Climate change is one of the most pressing challenges facing the world. It was first defined by the UN's Intergovernmental Panel on Climate Change (IPCC), which has been responsible for evaluating the science related to climate change since its establishment in 1988. The IPCC provides governments with scientific information about the causes of climate change, its impacts, and future risks in order to support the development of effective climate policies. According to the IPCC, climate change refers to "a change in the state of the climate that can be identified (e.g. using statistical tests) by a change in the mean and/or the ability of its properties, and that persists for an extended period, typically or longer." The IPCC finds that climate change can be caused by natural processes or by external forces (such as modulations to weather cycles, volcanic eruptions, and anthropogenic impacts on the atmosphere or land). The United Nations Framework Convention on Climate Change (UNFCCC), established in 1995, differentiates between climate change caused by human activities and natural climate variability. It defines climate change as "a change of climate which is attributed directly or indirectly to human activity that alters the composition of the global atmosphere and which is – in addition to natural climate variability – observed over comparable time periods." Addressing climate change by incorporating it into policy development and implementation, as well as accompanying measures, is crucial. The effects of climate change on ecosystems can be observed in the measurable warming of the atmosphere and oceans, the shrinking of glaciers, the break-up of ice on rivers and lakes, and rises in sea level. In addition, more frequent extreme weather events, including natural disasters, are increasingly causing damage to natural habitats and leading to the loss of biodiversity and other natural resources.

This chapter will explore the various effects of climate change on women at the environmental, economic, and social levels. It will also explain how gender has been integrated into the 2030 Agenda for Sustainable Development, as well as UNFCCC interventions through the

Lima Work Program, the Paris Agreement, and the Gender Action Plan. Additionally, it will provide an overview of the role of women in combating climate change on the political, technical, and scientific levels.

4.2 Vulnerability of Women to Climate Change

According to Robert Watson, chair of the IPCC, climate "vulnerability" can be defined as "*the extent to which the natural or social system is susceptible to sustaining damage from climate change and is a function of the magnitude of climate change, the sensitivity of the system to changes in climate. Hence, a highly vulnerable system is one that is highly sensitive to modest changes and one for which the ability to adapt is severely constrained*" (Olmos 2001). Women's vulnerability to climate change is influenced by various social, economic, and cultural factors. For example, 70% of the 1.3 billion people living in poverty are women and 40% of the poorest urban households are headed by women. In addition, although women produce the majority of the world's food (between 50 and 80% depending on the source), they own less than 10% of the land (Diagne Gueye 2008). Vulnerability to the impacts of climate change varies greatly between different regions and demographic groups, with significant consequences for marginalized segments of society, such as women, children, people with disabilities, and indigenous peoples. According to IPCC, the impact of climate change on individuals will change over time and depend on their ability to mitigate or adapt to change. Despite this, women often have less access to economic, political, and legal resources, making them even more vulnerable to the negative effects of climate change which can exacerbate existing gender inequalities.

In many cultures, the effects of climate change have distinct impacts on men and women. Typically, women are responsible for tasks such as producing and harvesting food, collecting water, and providing fuel for heating and cooking. Climate change can have a significant impact on these tasks, as reported by UN Women in 2022. Women have a crucial role in preserving nature and combating climate change, but limited land rights, as well as limited access to financial resources, training, technology, and political decision-making. These factors reduce women's ability to effectively contribute to climate change mitigations. Climate change, in short, has a different impact on men and women as a result of their gender-differentiated roles, powers, and responsibilities at the household and community levels.

Food Security

Women's food security (i.e., the ability to access sufficient, safe, and nutritious food to meet dietary needs) is crucial for women's well-being as well as the health of their families. When women experience food insecurity, it affects the entire family (Yadegari et al. 2017). As primary caregivers, women play a central role in providing and preparing food. When they struggle to provide nutritious meals, the health of all household members, especially the most vulnerable (children, the elderly, the sick, etc.), is compromised (Chilton et al. 2007). Women's food security is indeed intricately linked to maternal and child health. For example, inadequate nutrition during pregnancy and breastfeeding increases the risk of low-birth weights in babies and stunted growth in children, with long-term consequences for their development. Since access to resources (like income, land, and livestock) is vital for women's food security, social, economic, or cultural barriers that hinder their ability to produce or acquire food, perpetuate a cycle of poverty and vulnerability. Globally, women in Southern Africa are the most affected by these issues (WFP 2022).

According to FAO (2015), climate change has a significant impact on women's food security and their access to land and livestock. The effects of climate change—such as rising temperatures, changing precipitation patterns, and extreme weather events—directly affect agricultural productivity and the availability of natural resources, leading to substantial consequences for women's food security. Access to land is a critical factor for agricultural production and food security. Yet climate change exacerbates existing challenges and inequalities in land ownership and tenure, with women often facing greater disadvantages. In many regions—such as Africa, Asia, and Latin America—cultural norms and discriminatory practices limit women's land rights, making it difficult for them to access and control land resources (FAO 2018). Climate change-induced factors like droughts, floods, and soil degradation further compound these challenges. Women are disproportionately affected by land degradation, as they often rely on marginal and vulnerable agricultural lands, and also lack access to smart tools and technologies that can help mitigate climate change impacts. Women's limited access to credit and agricultural inputs also restricts their ability to adapt to climate change impacts. McKune et al. (2015) also note that although livestock provides valuable nutrition and income for many communities, extreme weather events, changing grazing patterns, and the spread of diseases affect livestock health and productivity.

Women, particularly in rural areas, often bear the responsibility for livestock rearing and management but lack knowledge and know-how solutions to overcome the impact of climate change on their animals. This can result in livestock losses, reduced income, and diminished access to animal-based food products, hence further jeopardizing women's food security. Climate change significantly impacts women's overall food security, yet the challenges and experiences of women differ significantly between low- and middle-income countries (LMICs) and developed countries. In LMICs, women play crucial roles in agriculture and face challenges such as water scarcity, limited access to modern farming technologies, and inadequate infrastructure. Climate change-induced disruptions affect food availability and increase food prices and make it harder for women to provide nutritionally adequate meals for their families, especially when and where farming provides the majority of household incomes. Limited access to education, healthcare, and diverse income-generating opportunities further hampers their ability to cope and adapt to climate change impacts. In developed countries, women have better access to resources, technology, and education, yet gender disparities still persist and effect women even in the most developed countries. Climate change disrupts agricultural systems, leading to food shortages, price fluctuations, and challenges in food supply chains, which particularly affect vulnerable groups like low-income households, single mothers, and elderly women (Nelson et al. 2010).

Health and Environmental Impacts

Climate change impacts—including more frequent and severe floods, droughts, and cyclones—disproportionately affect women in various countries. For instance, Niger, Senegal, and the Lake Chad basins have seen a 40–60% decrease in water availability (Odigie-Emmanuel 2010). Water shortages exacerbate health and sanitation issues, particularly in rural areas of Africa where access to water is already limited and create particularly negative impacts for women who tend to be tasked with carrying water to the home. During the Asian tsunami of 2004, an estimated 70% of the victims were women. This has been attributed to cultural and religious norms that kept women and children in their homes while men were out, confirming Wong's research that societal roles can restrict women's ability to make quick decisions in emergencies like natural disasters (Wong 2016). Climate change also impacts habitats, wildlife, ecosystems, and industries that rely on natural resources. Piao et al. (2010) assert that climate change will result in severe adverse changes to soils, arid lands, coastal zones, and tropical/boreal forests and severe treats to wetlands and vulnerable species. Rivers and dams are currently being used as dumping grounds for agricultural and industrial waste, which degrades primary water sources and impacts WASH activities, particularly for women and children. In West Africa, changes in weather patterns and climate conditions force people to use water from dirty sources, increasing their risk of waterborne illnesses. Furthermore, young girls are missing school as they spend more time fetching water and assisting with household tasks (Denton 2002).

Socioeconomic Impacts

Disparities between men and women's economic opportunities and access to productive resources are exacerbated by climate change. Women are often more vulnerable due in part to greater poverty, lack of education, and lack of representation in political, policy, and decision-making processes. In developing countries, women typically lack access to high-level positions and rely heavily on natural resources. A study by the World Bank (2015) revealed that 155 out of 173 economies have laws that impede gender equality. For example, in the agricultural sector, women constitute over 40% of the workforce but own only 10–20% of the land (varies from 20% in Latin America to 50% or more in some parts of Africa and Asia). Poverty and socioeconomic marginalization further weaken women's ability to cope with the challenges posed by climate change.

Natural disasters—such as droughts and floods—force women to work harder to sustain their households, limiting their opportunities for training, education, skills development, and income. Low-income groups in urban areas are particularly vulnerable to the effects of rising food prices caused by climate change. Hertel et al. (2010), for example, found that agricultural production is likely to be heavily impacted by climate change, with an associated 6.3% increase in the cost of living for those living in poverty. According to Gilbert and Morgan (2010), if global warming reaches 4 °C, cereal prices could increase by over 160% by 2080. This would be devastating for poor communities in developing countries, where around 70–80% of household income is spent on food and is likely to exacerbate hunger and poor nutrition. Food poverty disproportionately affects women, since they often give their own food to their children or prioritize high-quality food for their husbands and other family members. Poor female nutrition, as is well-studied, affects the health of not only women and mothers but also their children. Additionally, rising food prices may force women to leave school and seek employment (Ruel et al. 2010). Lagi et al. (2011) posit that high food prices can lead to an increase in societal issues, and research by Sternberg (2012) backs this up, linking the outbreak of violence in Egypt in 2011 to a food crisis caused by extreme weather conditions in other regions.

Climate change also affects human health, influencing incomes, economic development, education, social norms, migration, and access to healthcare, with particularly negative impacts on those living in poverty or experiencing social exclusion (Costello et al. 2009). Brody et al. (2008) demonstrated a direct correlation between climate change and rising health disparities, with climate impacts worsening the health of individuals living in poverty or those with pre-existing health conditions. Women and girls are particularly vulnerable to health issues as they are often responsible for caring for the sick, which increases their risk of infection and lowers their opportunities for generating income (Budlender and Moussie 2013).

4.3 Climate Engagement

The UNFCCC has made significant progress toward incorporating gender considerations into all aspects of its negotiations, despite not initially having a mandate on women's rights and gender equality. In 2014, the Lima Work Programme on Gender was established, and the Paris Agreement of 2015 acknowledged the importance of gender equality as a fundamental principle of climate action. In 2017, the Gender Action Plan (GAP) was also adopted. These efforts aim to enhance understanding and integration of gender equality in climate action initiatives.

Lima Work Program on Gender

In 2014, the first Lima Work Program was established at the twentieth session of the Conference of the Parties (COP 20) to further progress gender issues and (later) to integrate gender considerations into the Paris Agreement so as to achieve gender-responsive climate policy and action:

> *Parties to the UNFCCC have recognized the importance of involving women and men equally in the UNFCCC processes and in the development and the*

implementation of climate policies that are gender-responsive by establishing a dedicated agenda item under the Convention addressing issues of gender and climate change (UNFCC 2019). The Lima Work Program has been expanded and updated three times since 2014.

Paris Agreement

The Paris Agreement, adopted in 2015 at COP21, was an important step toward integrating gender in the UNFCCC. For example, the preamble to the Agreement states:

> *Acknowledging that climate change is a common concern of humankind, Parties should, when taking action to address climate change, respect, promote and consider their respective obligations on human rights, the right to health, the rights of indigenous peoples, local communities, migrants, children, persons with disabilities and people in vulnerable situations and the right to development, as well as gender equality, empowerment of women and intergenerational equity.*

Article 7 of the Paris Agreement further states that "*adaptation action should follow a country-driven, gender-responsive, participatory and fully transparent approach,*" while Article 11 advises that: "*Capacity-building should be guided by lessons learned, including those from capacity building activities under the Convention, and should be an effective, iterative process that is participatory, cross-cutting and gender-responsive.*"

Gender Action Plan

The Gender Action Plan was adopted in 2018 at COP 23 in Bonn. It "*seeks to advance women's full, equal and meaningful participation and promote gender-responsive climate policy and the mainstreaming of a gender perspective in the implementation of the Convention and the work of Parties, the secretariat, UN entities and all stakeholders at all levels.*"

The Plan has five priority areas:

- **Capacity-building, knowledge sharing, and communication** to enhance the understanding and expertise of stakeholders on the mainstreaming of gender considerations in policies, programs, and projects on the ground.

- **Gender balance, participation, and women's leadership** to achieve and sustain the full, equal, and meaningful participation of women in the UNFCCC process.
- **Coherence**—to strengthen the integration of gender considerations within the work of UNFCCC bodies, the secretariat, other UN entities, and stakeholders, and to consistently implement gender-related mandates and activities.
- **Gender-responsive implementation** to ensure the respect, promotion, and consideration of gender equality, and the empowerment of women in the implementation of the Convention and the Paris Agreement.
- Improved monitoring and tracking to strengthen implementation of—and reporting on—gender-related mandates under the UNFCCC.

4.4 Discussion of Women's Role in Response to Climate Change

Political Level

Policies and interventions that take a gender-based approach are more likely to have a sustainable and positive impact on communities. Women, therefore, have a crucial role to play in developing and implementing climate policies at the international, national, and local levels. However, despite their important role and positive impact on climate action, women are still marginalized and underrepresented in decision-making processes globally.

Practical and Technical Level

Women can play a critical role in planning and executing climate action initiatives at the international, national, and local levels when they have access to resources and information. Their unique knowledge and experiences can help develop and implement diverse and innovative solutions to the challenges of climate change. Women are active participants in informal reproductive and productive work (such as household/community care, seed/soil management, preservation of traditional agricultural knowledge, natural resource management of fuelwood, water, etc.). They also tend to be key decision-makers on issues impacting energy efficiency and household consumption (i.e., in the selection, use, and disposal of household goods and appliances).

Scientific Level

Several studies have highlighted the persistent gender inequalities in science and investigated the reasons for women's underrepresentation in high-level research (Hill et al. 2010; Long and Fox 1995; Pearson et al. 2015; White 1970; Settles et al. 2006;Moss-Racusin et al. 2012). A 2013 report in *Nature* found that despite some progress, female scientists still experience discrimination, pay disparities, and funding gaps, and that globally 70% of both men and women perceive science as a male-dominated field (Shen 2013). Women are also more likely to be professionally impeded by their caregiving responsibilities and are underrepresented in publications and citations (Budden et al. 2008; Dutt et al. 2016; Ceci and Williams 2011; Parker and Wang 2013).

Women's participation in research teams improves innovation and discovery (Nielsen et al. 2017; Loder 1999; Campbell et al. 2013). Several studies have explored the obstacles that women face in physical geographies, including the culture of masculinity in fieldwork, an absence of female role models, and the burdens of women's caregiving responsibilities (Maddrell et al. 2016; Luzzadder-Beach and Macfarlane 2000; Madge and Bee 1999; Bracken and Mawdsley 2004; Bee et al. 1998).

Research shows that the number of women participating in the Intergovernmental Panel on Climate Change (IPCC) has increased over time. In 1990, in the IPCC's first assessment report, there were no more than a dozen female authors and contributors (just 2% of all scientists involved). Numbers have subsequently increased, and 26 women contributed to the second assessment (1997), nearly 100 in the fourth report (2001) and 182 in the fifth report (2013). Only a few women held leadership roles in the IPCC until the sixth assessment (2023), but eight women are now in executive roles. (According to Miriam and Liverman (2018), however, women have had slightly more representation in the technical support units and task forces that support the IPCC.)

Despite societal structures that have obscured women's contributions to science, by the latter half of the nineteenth century, women were increasingly active in various scientific fields, including climate change; a pressing issue that the scientific community faces in the twenty-first century. Women today are actively engaged in the fight against global warming, and working to create a more sustainable relationship with the planet. Despite facing inequalities, women have shown their dedication and effectiveness in addressing climate change solutions and adaptations.

4.5 Summarizing the UN Recommendations: The Urgent Need for a Gendered Approach

The UN recommends that countries and policymakers increase their efforts to address climate change impacts by promoting gender equality (SDG 5) and urgent climate action (SDG 13). Countries should develop and implement projects that are led by women, and that specifically target women, especially those most vulnerable to climate change. It is crucial to support the implementation of the Paris Agreement by adopting gender equality guidelines in climate negotiations, ensuring women's representation, considering gender-specific impacts, and addressing the specific needs and rights of women. Capacity-building, technical assistance, and financial support should be provided to facilitate women's participation in climate negotiations, while a gendered approach should be applied across climate-related mechanisms and national documents.

Countries are encouraged to establish gender focal points. It is vital to report regularly on women's vulnerability and to ensure equal access—to opportunities, resources, decision-making, information, finance, technology, and knowledge—through capacity-building cycles. Thus, women's active participation in climate policy development, project implementation, knowledge exchange, and leveraging scientific research results for innovative solutions are strongly emphasized.

4.6 How Green and Digital Tools Can Support Climate Change Mitigation for Women

Green and smart technologies offer practical solutions to support women to mitigate the challenges of climate change. Enhanced access to digital information and smart tools enables women to avoid and reduce the negative impacts of climate change. Smart technologies can help promote sustainable agricultural practices, expand educational opportunities, and improve access to clean water and reliable energy sources. Sustainable farming practices and precision agricultural techniques enable women to adapt to climate variability and optimize agricultural production. For example, drip irrigation systems that deliver water directly to plant roots can minimize water wastage and enhance crop yields, while agroforestry—another climate-smart technique that combines trees with crops or livestock—can enhance soil fertility, conserve water, and provide additional income for women farmers.

Women's mobility, as well as their access to education and capacity-building programs, can be affected by climate change. Smart technologies address this by providing women with increased access to learning opportunities. Online platforms and mobile applications offer flexible and affordable education options, allowing women to pursue formal education or vocational training remotely. Massive Open Online Courses (MOOCs) offer free or low-cost courses on various subjects, empowering women to acquire new skills and qualifications that enhance their employability. Mobile learning apps provide women with access to educational resources anytime and anywhere, even in areas with limited infrastructure. Green technologies also have a direct impact on women's access to water and electricity. Water harvesting systems—such as rainwater collection tanks or rooftop water catchments—provide communities with a local and dependable water source. This eliminates the need for women to travel long distances to fetch water, reducing their workload, and allowing them to allocate time to other activities. Solar-powered water pumps eliminate the need for fuel and ensure reliable access to clean water for irrigation, household use, and livestock. Energy-efficient appliances, like cookstoves and LED lighting, decrease energy consumption, benefiting women who are responsible for household chores and managing household budgets.

4.7 Concluding Remarks and Recommendations

This chapter has shown that despite their heightened vulnerability, women can play a vital role in climate action. This includes developing climate policies, planning and implementing climate projects, and identifying innovative solutions through scientific research. The main challenge now is to systematically integrate (mainstream) gender equality and women's empowerment in our operational responses to climate change at the local, national, and international levels. To date, the international community has been slow to adopt and implement a gender-specific climate policy. A paradigm shift is now needed to ensure women are prioritized across all areas and levels. To do that we will need to adopt a gender equality strategy that involves men and women equally to design, implement, monitor, and evaluate policies and programs across the political, economic, legal, and societal spheres.

References

Bee, A., C. Madge, and J. Wellens. 1998. Women, gender, feminisms: Visiting physical geography. *Area* 30: 195–196. https://www.academia.edu/download/51204757/j.1475-4762.1998.tb00063.x20170105-14533-5drub.pdf.

Bracken, L., and E. Mawdsley. 2004. Muddy glee: Rounding out the picture of women and physical geography fieldwork. *Area* 36 (3): 280–286. https://www.jstor.org/stable/20004393.

Brody A, Demetriades J, Esplen E, Britain G. 2008. "Gender and climate change: Mapping the linkages, a scoping study on knowledge and gaps". BRIDGE Institute of Development Studies, University of Sussex. http://www.adequations.org/IMG/pdf/GenderAndClimateChange.pdf.

Budden, A.E., et al. 2008. Double-blind review favours increased representation of female authors. *Trends in Ecology & Evolution* 23: 4–6. https://journals.plos.org/plosone/article?id=10.1371/journal.pone.0003202.

Budlender D, Moussie R. 2013. Making care visible: Women's unpaid care work in Nepal, Nigeria, Uganda and Kenya. Action Aid, Johannesburg. https://actionaid.ie/wp-content/uploads/2016/10/Making-Care-Visible.pdf.

Campbell, L.G., S. Mehtani, M.E. Dozier, and J. Rinehart. 2013. Gender-heterogeneous working groups produce higher quality science. *PLoS One* 8: e79147. https://journals.plos.org/plosone/article?id=10.1371/journal.pone.0079147.

Ceci, S.J., and W.M. Williams. 2011. Understanding current causes of women's underrepresentation in science. *Proceedings of the National Academy of Sciences USA* 108: 3157–3162. https://www.pnas.org/doi/10.1073/pnas.1014871108.

Chilton, Mariana, Michelle Chyatte, and Jennifer Breaux. 2007. The negative effects of poverty & food insecurity on child development. *Indian Journal of Medical Research* 126 (4): 262–272. https://journals.lww.com/ijmr/Abstract/2007/26040/The_negative_effects_of_poverty___food_insecurity.3.aspx.

Costello, A., M. Abbas, A. Allen, S. Ball, S. Bell, R. Bellamy, S. Friel, N. Groce, A. Johnson, M. Kett, M. Lee, C. Levy, M. Maslin, D. McCoy, B. McGuire, H. Montgomery, D. Napler, C. Pagel, J. Patel, J.A.P. de Oliveira, N. Redclift, J. Rees, D. Rogger, J. Scott, J. Stephenson, J. Twigg, J. Wolff, and C. Patterson. 2009. Managing the health effects of climate change. *Lancet* 373: 1693–1733. https://doi.org/10.1016/S0140-6736(09),60935-1.

Fatma Denton. 2002. Climate change vulnerability, impacts, and adaptation: why does gender matter? *Gender and Development,* 10(2), Climate Change pp. 10–20. https://www.jstor.org/stable/4030569

Diagne Gueye, Yacine. 2008. Genre Changement Climatiques et Sécurité Humaine: Le Cas du Sénégal. (Dakar, Sénégal: Enda Programme Energie). https://studylibfr.com/doc/4149556/genre%2D%2Dchangements-climatiques-et-s%C3%A9curit%C3%A9-humaine

Dutt, K., D.L. Pfaff, A.F. Bernstein, J.S. Dillard, and C.J. Block. 2016. Gender differences in recommendation letters for postdoctoral fellowships in geoscience. *Nature Geoscience* 9: 805–808. https://www.semanticscholar.org/paper/Gender-differences-in-recommendation-letters-for-in-Dutt-Pfaff/75fa1395ce8e2de930064c20074d93593ebb11d3.

Food and Agriculture Organisation—United Nations. 2015. Climate change and food security: risks and responses. https://www.fao.org/3/i5188e/I5188E.pdf

———. 2018. Gender gap in land rights. https://www.fao.org/3/I8796EN/i8796en.pdf

Gilbert, C.L., and C.W. Morgan. 2010. Food price volatility. *Philosophical Transactions of the Royal Society* 365: 3023–3034. https://doi.org/10.1098/rstb.2010.0139.

Hertel, T.W., M.B. Burke, and D.B. Lobell. 2010. The poverty implications of climate-induced crop yield changes by 2030. *Global Environmental Change* 20 (4): 577–585. https://doi.org/10.1016/j.gloenvcha.2010.07.001.

Hill, C., Corbett, C., and St Rose, A. 2010. Why so few? Women in science, technology, engineering, and mathematics, American Association of University Women. Washington, DC. https://eric.ed.gov/?id=ED509653.

Lagi M, Bertrand KZ, Bar-yam Y. 2011. The food crises and political instability in North Africa and the Middle East. New England Complex Systems Institute, Cambridge. https://www.academia.edu/833188/The_Food_Crises_and_Political_Instability_in_North_Africa_and_the_Middle_East.

Loder, N. 1999. Gender discrimination undermines science. *Nature* 402: 337. https://www.nature.com/articles/46379.

Long, J.S., and M.F. Fox. 1995. Scientific careers: Universalism and particularism. *Annual Review of Sociology* 21: 45–71. https://www.jstor.org/stable/2083403.

Luzzadder-Beach, S., and A. Macfarlane. 2000. The environment of gender and science: Status and perspectives of women and men in physical geography. *The Professional Geographer* 52: 407–424. https://onlinelibrary.wiley.com/doi/abs/10.1111/0033-0124.00235.

Maddrell, A., K. Strauss, N.J. Thomas, and S. Wyse. 2016. Mind the gap: Gender disparities still to be addressed in UK higher education geography. *Area* 48 (1): 48–56. https://www.jstor.org/stable/24812194.

Madge, Clare, and Anna Bee. 1999. Women, science, and identity: Interviews with female physical geographers. *Area* 31 (4): 335–348. https://www.ncbi.nlm.nih.gov/pmc/articles/PMC5834669/.

McKune, Sarah L., Erica C. Borresen, Alyson G. Young, Thérèse D. Auria, Sandra L. Ryley, Astou Diao Russo, Meghan Coleman Camara, and Elizabeth P. Ryan. 2015. Climate change through a gendered lens: Examining livestock holder food security. *Global Food Security.* 6: 1. https://doi.org/10.1016/j.gfs.2015.05.001.

Miriam, Gay-Antaki, and Diana Liverman. 2018. Climate for women in climate science: Women scientists and the intergovernmental panel on climate change. *PNAS* 115 (9): 2060–2065. https://www.pnas.org/doi/10.1073/pnas.1710271115.

Moss-Racusin, C.A., J.F. Dovidio, V.L. Brescoll, M.J. Graham, and J. Handelsman. 2012. Science faculty's subtle gender biases favour male students. *Proceedings of the National Academy of Sciences USA* 109: 16474–16479. https://www.pnas.org/doi/10.1073/pnas.1211286109.

Gerald C. Nelson, Mark W. Rosegrant, Amanda Palazzo, Ian Gray, Christina Ingersoll, Richard Robertson, Simla Tokgoz, Tingju Zhu, Timothy B. Sulser, Claudia Ringler, Siwa Msangi, and Liangzhi You. 2010. Food security, farming, and climate change to 2050: Scenarios, results, policy options. Illustrée. Intl Food Policy Res Inst. isbn 0896291863, 9780896291867.

Nielsen, W., et al. 2017. Correction for Nielsen et al., opinion: Gender diversity leads to better science. *Proceedings of the National Academy of Sciences USA* 114: E2796. https://www.ncbi.nlm.nih.gov/pmc/articles/PMC5380096/.

Odigie-Emmanuel, O. 2010. The gendered impact of climate change in Nigeria. In: Dankelman, I. (Ed.), Gender and climate change: An introduction. Earthscan, London, pp. 123e129. https://ps.booksc.me/book/66610400/6872ef

Olmos, S. 2001. Vulnerability and adaptation to climate change: Concepts, issues, assessment method paper. Climate Change Knowledge Network Foundation Paper. https://docplayer.org/amp/73034834-Die-genderbilanz-des-klimadiskurses.html.

Parker, K., and Wang, W. 2013. Chapter 6: Time in work and leisure, patterns by gender and family structure. Pew Research Center. https://www.pewresearch.org/social-trends/2013/03/14/chapter-6-time-in-work-and-leisure-patterns-by-gender-and-family-structure/

Pearson, W.J., Frehill, L.M., McNeely, C.L. 2015. Advancing women in science and engineering. eds Pearson WJ, Frehill LM, McNeely CL, Berlin: Springer. https://www.amazon.com/Advancing-Women-Science-International-Perspective-ebook/dp/B00WLMF3P2.

Piao, S., P. Ciais, Y. Huang, Z. Shen, S. Peng, J. Li, L. Zhou, H. Liu, Y. Ma, Y. Ding, P. Friedlingstein, C. Liu, K. Tan, Y. Yu, T. Zhang, and J. Fang. 2010.

The impacts of climate change on water resources and agriculture in China. *Nature* 467: 43e51. https://pubmed.ncbi.nlm.nih.gov/20811450/.
Ruel, M.T., J.L. Garrett, C. Hawkes, and M.J. Cohen. 2010. The food, fuel, and financial crises affect the urban and rural poor disproportionately: A review of the evidence. *Journal of Nutrition* 140 (1): 1705–1765. https://pubmed.ncbi.nlm.nih.gov/19939990/.
Settles, I.H., L.M. Cortina, J. Malley, and A.J. Stewart. 2006. The climate for women in academic science: The good, the bad, and the changeable. *Psychology of Women Quarterly* 30: 47–58. https://journals.sagepub.com/doi/10.1111/j.1471-6402.2006.00261.x.
Shen, H. 2013. Women in science: Women's work. *Nature* 495: 21. https://www.pnas.org/doi/pdf/10.1073/pnas.1710271115.
Sternberg, T. 2012. Chinese drought, bread and the Arab spring. *Applied Geography* 34: 519–524. https://doi.org/10.1016/j.apgeog.2012.02.004.
UNFCC. 2019. Introduction to gender and climate change. https://unfccc.int/gender#:~:text=Parties%20to%20the%20UNFCCC%20have,and%20climate%20change%20and%20by
White, M.S. 1970. Psychological and social barriers to women in science. *Science* 170: 413–416. https://www.science.org/doi/10.1126/science.170.3956.413.
Wong, S. 2016. Can climate finance contribute to gender equity in developing countries? *Journal of International Development* 28 (3): 428–444. Institute of Development Studies https://onlinelibrary.wiley.com/doi/full/10.1002/jid.3212.
World Bank Group. 2015. Women, business and the law 2016: Getting to equal. https://openknowledge.worldbank.org/handle/10986/22546.
World Food Programme. 2022. Climate crisis threatens food security for women and girls. https://www.wfp.org/news/climate-crisis-threatens-food-security-women-and-girls.
Yadegari, L., M. Dolatian, Z. Mahmoodi, S. Shahsavari, and N. Sharifi. 2017. The relationship between socioeconomic factors and food security in pregnant women. *Shiraz E-Medical Journal* 18 (1): e41483. https://doi.org/10.17795/semj41483.

PART II

A Sectorial Vision

CHAPTER 5

Gendered Considerations in Relation to Climate Change and Agriculture: Reflections on Impact, Funding, and Mainstreaming

David Lefor and Sujata Ganguly

5.1 Introduction

Women have long played an integral—if often unrecognized—role in agriculture worldwide. Ester Boserup's influential book, ***Woman's Role in Economic Development,*** published in the early 1970s challenged the notion of women as passive beneficiaries of development (Boserup 1970). She called for a focus on Women in Development (WID), urging development

D. Lefor (✉)
Ruhr University Bochum, Bochum, Germany

Independent Researcher, Bangkok, Thailand
e-mail: David.Lefor@edu.ruhr-uni-bochum.de

S. Ganguly
Includovate, New Delhi, India
e-mail: sujata.ganguly@includovate.com

K. Drucza et al. (eds.), *Centering Gender in the Era of Digital and Green Transition*,
https://doi.org/10.1007/978-3-031-38211-6_5

practitioners to acknowledge women's often invisible labor, to pay attention to the specific needs and preferences of women when designing and implementing development projects, and to champion a more general understanding and appreciation of the critical roles women play in agriculture and rural development.

At the first United Nations' Conference of the International Women's Year in Mexico City in 1975, there was much more focus on the roles women play in effecting rural transformation in the global south. The Women in Development movement was ultimately succeeded by the Gender and Development (GAD) movement in the 1980s. This recognized that understanding gender roles and relations is key to improving women's lives. With the introduction of the term 'gender', the movement ushered in a new interest in understanding both women and men's need and a deeper focus on addressing unequal relations and inequities. At the 1995 Fourth World Conference on Women in Beijing, perhaps the most influential conference to date to consider gendered issues, state participants committed themselves (through the Beijing Platform for Action) to establishing mechanisms to promote women's rights—including national action plans, gender strategies, and legal frameworks. In 2000, states confirmed their commitment to reducing gender inequalities through the United Nations Millennium Declaration. These successive milestones for the international community led to an increase in donor interest and subsequent funding for gender-focused programs. Building upon these milestones, the State of Food and Agriculture 2010–11 brought high-level attention to gender in agriculture, and the 2012 World Development Report further raised the profile of gender by indicating significant ways in which it could be integrated into development programs.

Despite these achievements, significant challenges remain to be solved or answered. Nowhere in the world are women born into full equality and, in many places, to be born female is to be born a second-class citizen (Donor Tracker 2020). In many parts of the world, women are less likely to be in positions of power, own land, have reproductive rights, gain education, and have access to the job market. Women and girls suffer disproportionately as a result of worsening droughts, floods, fires, and storms, and are at a higher risk of displacement. In many communities, women work in the roles and sectors that are most susceptible to climate change, threatening their security and livelihoods (Women in Climate Action Group 2021).

The past decade has increased international attention on gender equality in agriculture and development. This led to policies calling for an increased focus on gender equality issues, as well as for feminist aid and foreign policy focusing on gender. One of the largest challenges in monitoring gender equality commitments is tracking the funding and programming globally (Grabowski and Essick 2020). The agriculture sector has a remarkably high share of aid allocated for gender equality and women's empowerment compared to other development sectors, with around 60 percent of agricultural programs having gender as a focus area in recent years (OECD DAC Network on Gender Equality 2020).

> *Of all economic and productive sectors, agriculture receives by far the most gender-related funding; it attracted over one-quarter of all gender-related funding, or US$3.4 billion in 2018. The agricultural sector has historically been viewed as a cornerstone of development efforts overall. Training and capacity building for small farmers can serve many development-friendly purposes: for example, increasing food production to alleviate hunger, building out local economies in rural communities, and providing jobs in farming and crop sales. Because of the great potential of agricultural development projects to deliver on multiple development goals at once, and because a large number of women are already involved in and employed in the agriculture sector, it is also an attractive target for gender-related spending.* (Donor Tracker 2020)

A gender lens is often not fully considered when it comes to project implementation and research especially pertaining to climate change and agriculture in Africa and Asia. Agricultural research for development must not shy away from a feminist, transformative agenda that questions and confronts power and politics—whether between women and men, within households, communities, and countries; or between different groups in a community (Pyburn and Eerdewijk 2021). After considering the history of feminist approaches and the extent to which donors, governments, and implementers are currently prioritizing and integrating a gender lens into interventions and research projects (especially relating to the impacts that climate change has on agriculture) this chapter explores the 'bigger picture' and articulates the next generation of transformative gender research.

5.2 Conceptual Framework

Gender equality and women's empowerment results are not easy to measure. Progress toward gender equality is also slow, multifaceted, and complicated, since it often includes influencing norms and attitudinal changes. Reflective and innovative approaches in terms of interventions, funding, and planning are therefore needed to bring about a transformational change. At the same time, donor agencies need to work to strengthen their capacity to manage development results and address gender equality and women's empowerment explicitly. However, although aid allocated for gender equality is increasing overall (according to the 2018 OECD DAC Network on Gender Equality), funding for dedicated gender equality programs is lagging behind. There is also a dearth of literature exploring the shifts in donor priorities considering the evolving landscape of feminist approaches within development.

This chapter also examines the opportunities that are available but yet to be tapped. That involves understanding the barriers and obstacles to carrying out gender research projects or implementing programs with a gendered component, especially those that seek to adapt or mitigate climate change impacts on agriculture in Asia and Africa. To do this, our key research questions are:

- How can gender mainstreaming be ensured in the development sector?
- What are donors' priorities?
- What are the specific challenges in terms of integrating a gender lens into programmatic interventions (specifically on climate change)?
- What are the untapped opportunities available to address gender equity in climate change research for development?

5.3 Methodology

Our research followed a qualitative approach. An extensive (traditional) desk review was followed by key informant interviews. The key informants were selected based on purposive sampling from different locations (Asia, Africa, and North America) and different professions (practitioners, researchers, and academics). In total, information was collected from 20 key informants, all of whom are experts working on gender, climate change impacts, and agriculture. Prior to data collection, the research was submitted to an institutional review board (IRB) and obtained ethical clearance.

5.4 Findings and Discussions

Rethinking Gender in Agriculture

As has been underscored by the extensive and growing body of literature on gender and agriculture, women's roles and responsibilities in agriculture—both within the household and the community—are complex, diverse, and multifaceted (Doss 2001; Doss and Morris 2001; Meinzen-Dick et al. 2011; Quisumbing and Pandolfelli 2010). Women play a significant role in agriculture and its allied fields such as crop production, livestock production, horticulture, post-harvest operations, agro/social forestry and fisheries, and so on. Studies suggest that, despite the wide variability between regions and countries, women farmers play a major role in labor-intensive agricultural activities such as planting and weeding (Meinzen-Dick et al. 2011). Despite this, the title of 'farmer' is culturally and normatively ingrained into most societies as a typically male profession (just as it is common to consider 'nursing' a female profession). Women farmers, therefore, are often considered to be 'farmers' wives' or just 'helpers'. As we move forward, exploring the roles of women and their identities in the context of agriculture, it is critical to move away from labels like 'farm women', and 'farmers' wives' (except where that is simply the case), to identifying women as 'farmers' in their own right, and in acknowledgment of their significant contributions to agriculture. As Ester Boserup remarked, '*economic development could not be fully evaluated without the recognition of the innumerable "hidden contributions" of women throughout the world, particularly in the form of unpaid work*' (Quisumbing Agnes et al. 2014).

Men continue to dominate economic and household decision-making even in areas where women provide the main source of labor, in part because the influence of women has not been recognized. When scouring through current literature, however, it is noticeable how we have tried to define women's roles, responsibilities, and involvement in agriculture. The literature discusses a triple burden on women as a result of the impact of patriarchy (especially 'classic patriarchy'[1]), failures to address intersectionality, and failures to acknowledge agency and the full extent of women's involvement in agriculture. These factors have direct impacts on women's levels of land ownership and inheritance, the types of work available to them, their socioeconomic stratification, their ability to engage with—and own—technology, as well as the impacts that climate change brings. One key question to consider is: How can we create development programs that take a transformative approach throughout design and implementation, and which are able to address the norms which foster inequity? One of our key informants, a Senior Agronomist from India, recognizing the importance of such considerations, stated:

> *I recognize that gender-inclusive outcomes can be realized only when women are mainstreamed into the design and delivery of technology and knowledge from the beginning, and this requires a greater commitment to equity and inclusion. The need for technological and institutional interventions that foster gender equity and create opportunities for female and male farmers is key to sustainable, resilient, and remunerative agriculture in regions like eastern India where we work.*

In transformative programs, gender mainstreaming considers the factors that can aggravate vulnerabilities. It does not focus on gender merely for the sake of it, but to ensure that there is learning and self-reflection on gendered norms, beliefs, assumptions, and unconscious biases—all of which can lead to greater inclusion.

> *Gender mainstreaming is being used as a blanket practice without understanding the nature of the programs.*—Practitioner, India

While gender mainstreaming is complex, offering both challenges and opportunities, it is extremely important in programming—as many respondents confirmed—and particularly in agriculture, to counter the increased vulnerabilities women face to climate change. To use one simple

[1] Kandiyoti, Deniz (1988) definition of 'classic patriarchy' details a patrilocal residence pattern, where women are expected to live with or near the husband's family and where male-dominated decision-making within households are the norm. Women expect, over time, to be able to exercise power based on their status as mothers of sons and mothers-in-law, a situation whereby older women have gains to make from the subordination of younger women.

example, by adjusting the designs of tractors we can allow them to be more easily driven by women, not only by men. In terms of self-realization and understanding, it is important to ask how effective the gender dimensions of an intervention are, how people on the ground view them, and how sustainable the gender-focused intervention may be (e.g. if and when a 'gender expert' leaves a project mid-way, ongoing support and budgeting for gender mainstreaming work can disappear). The extent to which there is gender 'follow through' depends on the capacity (and prioritization) of management, staff, and other project stakeholders, and their willingness to support gender mainstreaming efforts. It has been noted by the experts who were interviewed as part of this research that when gender mainstreaming is not incorporated at the beginning of programming, it is less meaningful and impactful. In addition, when work is completed in 'silos' (i.e. as a series of compartmentalized specializations of work instead of a mass coordinated effort) it can also reduce the effectiveness of gender mainstreaming.

The inclusion of women in capacity building and other opportunities also faces cultural and social barriers, particularly when women don't have equal access to services and knowledge due to their status in the family:

> *Most other projects are implemented by NGOs where focus is on classroom training or organic farming which are not leading to economic empowerment. There is no good method to quantify the impact of such training except the number of beneficiaries, which is not helping the cause of women farmers.*—Senior Agronomist, India

The gender gap has been successfully decreased when gender mainstreaming is incorporated throughout projects and programs, starting at conceptualization and moving through planning, recruitment, implementation, dissemination, and evaluation of activities. Organizations which mainstream a feminist approach and exhibit deep familiarity with gender policies and programming (as well as in literacy in gender mainstreaming research and theories of change) have found success in gender-transformative change. Ensuring that the right partnerships are in place—for example, through women-based collectives and livelihood organizations—also helps ensure that women are placed at the center of programming. With the renewed push to increase gender-focused work in agriculture and to mitigate the gendered impacts of climate change, we must ask: *What is holding back true transformational change?*

Many experts believe that although gender mainstreaming is an ideal to work toward, it is often watered down or not fully embedded in program approaches. Part of the problem is that most people don't *really* know what mainstreaming means, and typically operate within a traditional, singular approach based on existing norms thought to be optimal. Gender oppression is often the norm, not just for women but of all people who don't identify with a narrow range of gender identities that only privilege certain classes of individuals. Gender mainstreaming's pluralistic approach, which considers the diversity among genders, is a step in the right direction, but doesn't completely solve this kind of problem. Additionally, gender concepts around the world are often binary. It is quite rare, even in the twenty-first century, to consider a non-binary approach in the development sector. There are considerable differences in how societies and communities recognize gender and how people express their gender identities. Trying to combine all this diversity into a binary blurs the specific issues that different people face in different places. It is crucial to allow gender-diverse people to speak for themselves, to represent their own needs and preferences. This not only ensures that practitioners, policymakers, and researchers hear the full diversity of experience, but also gives gender-diverse people the power of self-representation, which they are often denied. It is worth noting that not all gender-diverse individuals want their identities called out in the form of targeted interventions. Talking to gender-diverse people is the only way to ensure that program interventions address the issues they face in ways they find respectful.

Rethinking Mainstreaming: Lessons from Gender Experts in Development Our research findings illuminate the following best practices/opportunities to rethink gender mainstreaming approaches:

Address social norms: Rethink whether research/programs address gender in a tokenistic way, or deep dive into social norms to address the root cause of problems.

Promote systemic change: Ensure access, participation, decision-making, and well-being.

Follow a bottom up 'transformative' approach: Involve end-users in concept and design phases to ensure the needs of diverse people are fully understood from the beginning.

Work together not in silos: Ensure key stakeholders share a common vision and implement together rather than in silos.

Generate evidence: Include the voices of marginalized farmers irrespective of gender identities and ensure the availability of robust data to support policymakers.

Avoid a binary approach: Ensure program interventions and research extend beyond the binary approach so that programs and research are inclusive and leave no one behind.

Funding Ecosystem

The funding ecosystem is one of the most under-researched topics when it comes to gender in agriculture and climate change. There is a need to understand donors' priorities and their shift in focus—both substantive and financial. In addition, considerations of the funding ecosystem tend to get diluted and diverted due to a focus on 'economic growth' or 'returns on investment' rather than a rights-based perspective.

Donor priorities tend to view gender as a 'cross-cutting issue' in development work. On the one hand, this means that gender has become an issue that can't be ignored, with many donors now requiring gender action plans/commitments toward gender equality. This is a good thing and clearly shows that the years of hard work by advocates, activists, and gender specialists have paid off. On the other hand, the way in which 'gender' has traditionally been operationalized in research is often simplistic, binary, and un- or under-contextualized. Currently there are varying stances on what applying and implementing a gender analytic lens into projects actually looks like in practice. At one end, you have in-depth, intersectional gender analyses that views equity as central to informing a project's pathways to success. At the other end, you have 'tick the box' activities where data is collected to reflect project goals (e.g. the number of women trained in a new agricultural technology) without attention to the gender norms and intrahousehold/community dynamics that mediate women's access to, use of, and control over that technology. Ultimately, it really depends on whether a project team wants to pay real attention to gender norms (and work toward solutions that transform harmful norms and reinforce gender inequality) or whether they want to tick boxes to satisfy a donor requirement. According to one key informant:

> *Quite a few donors are chasing numbers, some others are really into this transformative approach. For example, just putting women as leaders for the sake of*

it kills inclusion. You need leaders, not just women. Enforcing norms/rules is not ideal.—Senior Agronomist, India

Two distinct schools of thought emerged from our key informant interviews. One group felt that donors have shifted their priorities toward gender mainstreaming, while another felt that donors lag behind and need conceptual shifts. The following section considers these two opinions in more detail.

Donor Priorities Have Shifted Toward Gender Mainstreaming (7 Out of 20 Key Informants) Some informants suggest that donors have become much more gender-sensitive and that prioritizing gender inclusion is 'the new normal'. This is a positive development both for donors and gender mainstreaming, highlighting that from the donors' side, there are higher chances of stakeholders being highly aware of the importance of gender work, and that programs cannot continue to adopt a gender-biased lens.

Recently our donors have focused on bringing a gender lens to our agronomy and soil-related work, and in new proposals gender is a cross cutting theme that focuses on greater integration at state-level of policy and recommendations across agriculture and livelihoods.—Researcher, India

Most of our donors do have gender equality and social inclusion (GESI) and human rights in their list. In most of the projects, except for some very technical ones, a gender lens can be applied. However, we do face problems and challenges from the technical professionals due to their limited understanding of a 'gender lens'. Often if there is a budget constraint, then the gender-related work gets dropped!—Gender specialist, Nepal

Donor Priorities Lag Behind and Require Conceptual Shifts to Become Effective (8 out of 20 Key Informants) Donor priorities appear to be increasingly short-term and mainly concerned with emergency relief rather than long-term, stable investment. This reflects fatigue with development aid among the voters in donor countries. The argument that rich countries are responsible for the fast-approaching, global climate change catastrophe and should therefore subsidize mitigation and adaptation for poor countries is not widely subscribed to among voters.[2]

[2] One key informant was able to comment from their expertise.

> *The rest of the world has to stop looking to the rich countries, the USA in particular, for leadership on gender-sensitive climate reform. This requires a revolution in collective action.*—Gender Advisor, USA

At the same time, wide-scale concern/awareness about climate change, natural disasters, and species extinction are growing as political issues, especially among the young who look at their world through a gender lens that is quite different from the one conventionally applied in development projects. The political environment in which donors operate therefore seems highly volatile and there is no evident commitment to action at sufficient scale or speed to avoid irreversible changes in which millions of people will lose their livelihoods and their lives.

> *The application of a gender lens to projects is widely institutionalized in the sense that donor-funded proposals and projects are expected to pay some attention to gender but few donors seem willing or able to challenge the power relations that underpin gender inequality in poor countries where projects are targeted.*—Gender Advisor, USA

> *I believe donors should value and pay close attention to the extent to which the issue of gender is considered both in the composition of the team that they are considering to award/contract any assignments related to climate change and conservation of ecosystems and in the research /evaluation criteria of those assignments.*—Senior Researcher, Ethiopia

In addition, gender transformation is foundationally interconnected to cultural issues of race, language, and community preferences and requires a shift in resources from centers of wealth to currently underserved regions. If Africa is going to be less poor, Europe and other developed countries have to be less rich.

> *As an American, I actually think our money doesn't buy us happiness. Less would be more but wealthy governments and especially foundations funded by billionaires, don't believe that and don't want to hear it.*—Professor, USA

It could be said that prioritizing gender inclusion is the new normal—a very positive development for both donors and gender mainstreaming. While an increased interest in gender exists, donors may not necessarily know or understand what a gender-responsive or gender-transformative project would look like in practice. Some experts consulted for this chapter note that there are still struggles in delivering transformative change through mainstreaming.

5.5 Challenges Integrating Gender into Research Projects and Programs

Climate change, lack of access to water, and losses in biodiversity affect people at all levels. That said, these impacts are unequal and intersectional. Identities (like gender and disadvantaged economic conditions) interact with systems of power to determine how people can adapt to these changes at the individual, household, community, and societal levels.

Lack of expertise and lack of financial resources can jeopardize the integration of gender-transformative lenses into programs and research projects. Prior efforts to empower women around the globe have sometimes failed because they have not fully considered the gendered context of women's lives, the intersectionality of discrimination against women, and the deeply ingrained nature of gender inequality at a structural and political level. Furthermore, capacity to design and implement research to effectively address gender equity is not level across institutes and programs. In most cases, the conceptual understanding stays with project leaders and those writing strategies and may not necessarily trickle down to other levels of the institute or members of the team. Considerations of gender concepts can then become diluted across various levels and parts of the project until gender becomes only a project formality.

> *Advancing women's equity in agriculture/water/biodiversity is in almost every political speech and policy document. However, it is difficult to implement these policies on the ground, and the reason I see for this situation is the lack of training and awareness of the stakeholders on the application of gender considerations in the design and implementation of agricultural projects.*—Senior Scientist, Nigeria

The most difficult challenge is to create the required changes at all levels and among all types of actors and organizations to institutionalize accountability for outcomes that improve gender equality.

> *No one is losing funding or getting fired for not delivering: there's a lot of talk but no one is held materially responsible if women aren't better off.*—Professor, USA

The monitoring and evaluation of programs often fails to fully consider and integrate a gender lens. Only recently did the OECD-DAC indicators take gender seriously, and it is still easy for organizations, in their reporting documents, to discuss various concepts around gender but fail to include any significant considerations for the implementation or monitoring of gender components. Of course, very often, gender has also not been included appropriately in program design. If donor reporting doesn't require adequate consideration (and evidence) for gender-related indicators, and if the interventions continue to exclude/ignore/downplay gender, then it will take a long time for change to come.

Differences in the definitions or understanding of women's empowerment is another challenge. For example, simply providing opportunities for women—such as training or capacity building programs—is not sufficient unless they are also able to take on decision-making roles and have decision-making power. Field implementers must be gender-sensitive and must understand the importance of positive discrimination for promoting women. Women need to be given added opportunities and a conducive environment to grow. At a policy and institutional-level, states need to avoid placing women in positions of power without ensuring that systems are in place to nurture their capacity to lead. Additionally, marginalized communities are not protected sufficiently from the potential negative impacts of the promotion of women's empowerment, and thus caution must be exercised when localization in planning takes place. For example, in biomass energy and cooking, preparing food is a deeply traditional activity with normative pressures on women to both harvest biomass for fuel and to prepare the food. Thus the development of normative shifts toward empowerment must proceed carefully, considering cultures, customs, and contexts. Outside sponsors tend to dismiss cultural preferences that they don't understand. Programs that aim to empower women farmers often show no difference in designing interventions for them and designing interventions for male farmers. Yet there can be wide differences in the levels of understanding between male and female farmers (due to education, exposure, social norms, etc.) which can affect even basic level trainings. Therefore, we have to purposefully design gender-specific programming until equity is reached, for example, by oversampling women farmers and ensuring greater participation in field days—support on this must be from the start and from the top. COVID-19 has exacerbated climate and resource crises, along with global inequalities. Today, climate plans, policies, and investments still do not adequately account for the distinct impacts of climate change on women, girls, and marginalized populations. Regardless of the sector, it is very often women and girls who are expected to carry the increased burden caused by climate change, as well as the pandemic (Hung et al. 2021). Again, this points to the importance of prioritizing gender-specific interventions which ensure that girls and women are not only able to participate, but can lead through empowerment and capacity building.

5.6 Untapped Opportunities

Despite the challenges in integrating a gender-transformative lens into programming, there are opportunities that can be tapped for impactful change. We can learn about these by examining best practices relayed from the field, including by the practitioners, researchers, and professors interviewed for this study. In India, one expert described how their organization has been dealing with women's unequal access to knowledge and technology through a livelihood-centric approach. The activities undertaken included formative research, as well as actions for catalyzing and promoting innovation, generating evidence, and collecting lessons learned to enable the mainstreaming of inclusive policy actions. These actions, from the ground up, consider women in every stage of the project.

> *We are engaging women in innovation in rice/wheat agriculture systems in eastern India through strategic partnerships and livelihood institutions. Realizing that women farmers have limited access to new knowledge, technologies, and information due to the challenges that traditional extension systems face in reaching out to women, our projects intervened with an alternative model of combining research and extension activities making it more gender inclusive. Following a livelihood-centric approach, the activities undertaken included formative research, action for catalyzing and promoting innovation processes, and generation of evidence and learning to enable mainstreaming of inclusive policy actions.*—Agronomist, India

Considering the untapped opportunities relating to women's issues and climate change impacts in agriculture, the research suggests the following key avenues to mainstream gender:

- Prioritize the voices of women, men, and non-binary people from the communities in which interventions are 'implemented'. They need to be heard and considered from day one.

- Understand how women and men (differently) perceive, face, and respond to climate change through inclusive gender-sensitive research that can catalyze the design of gender-sensitive, needs-based interventions.
- Develop longer-term indicators and metrics to measure project progress toward gender equality and other goals identified as relevant by local people.
- Utilize a more integrative approach to gender inclusion in terms of data capture, design, and implementation.
- Increase women's participation in programs, for example, by investigating whether decoupling agricultural assistance from land ownership improves participation.
- Expand women's entrepreneurship skills.
- Analyze the impact of technology developed considering gender-inclusive approaches.
- Increase gender inclusion through targeted innovations such as collective action (i.e. women's self-help groups) and gender-sensitive training and extension services.
- Develop institutional systems to integrate women in post-processing value chains.
- Develop behavioral change communication to reduce gender gaps in intrahousehold decision-making.
- Find ways to mobilize and empower women across social classes, races, ages, and national boundaries to demand climate reform.
- Ensure accurate information on climate change is integrated into popular culture to mobilize the public and foster grassroots change.

5.7 Concluding Remarks and Recommendations for Transformational Change

Gendered dimensions are often not fully considered when it comes to project implementation and research especially in climate change and agriculture. Researchers, policymakers, and development practitioners are increasingly acknowledging the gendered dimensions needed to effectively address climate variability and change within agriculture. However, the 'add women and stir approach' to gender, which is often used, fails to address gender gaps fully or in a holistic way.

Climate change has a disproportionate impact on women and girls, and failures to recognize or act on this reality can halt or even reverse progress toward gender equality. Although, encouragingly, gender is receiving more attention at the program level and overall funding for it has increased, budgetary allocations for specific gender equality programs are currently lagging behind. In addition, programs that do integrate gender often do so in a tokenistic way that only engender superficially results, but do not define and detail the critical pathways that can create systematic global long-term changes. We need to be intentional in addressing gender inequality and make women-focused investments to build their resilience to climate change.

Recommendations for Transformational Change:

- Promote grassroots, local governance, and community-led natural resource management
- Increase the numbers of women in STEM
- Fund more qualitative research
- Use participatory research methods and transformative gender approaches
- Fund women who are disproportionately affected by climate change to go to capital cities to speak to policymakers and to go to UN conventions internationally to have their voices heard
- Promote women as change agents, not victims of climate change
- Generate evidence to address gender equity at a program level
- Promote convergence between diverse partners and stakeholders to reduce siloed work and increase coherence
- Dedicate sufficient time and budget to integrating a gender lens into programs

References

Boserup, Ester. 1970. *Woman's role in economic development.* New York: St. Martin's Press.

Donor Tracker. 2020. Investing in shared prosperity: Financing for Women's economic empowerment. Donor Tracker Insights. Available at: https://donortracker.org/insights/investing-shared-prosperity-financing-womens-economic-empowerment.

Doss, Cheryl R. 2001. Designing agricultural technology for african women farmers: lessons from 25 years of experience. *World Development* 29 (12): 2075–2092.

Doss, Cheryl R., and Michael Morris. 2001. How does gender affect the adoption of agricultural innovations? The case of improved maize technology in Ghana. *Agricultural Economics* 25: 27–39.

Grabowski, Aria, and Parker Essick. 2020. Are they really gender equality projects? An examination of donors' gender-mainstreamed and gender-equality focused projects to assess the quality of gender-marked projects. Oxfam. 10.21201/2020.5655.

Hung, Samantha, Mohammad Naciri, and Sun-Ah Kim. 2021. Can COP26 become a key moment for gender responsive climate action? https://www.adb.org/news/op-ed/can-cop26-become-key-moment-gender-responsive-climate-action-samantha-hung-mohammad.

Kandiyoti, D. 1988. "Bargaining with Patriarchy." *Gender and Society* 2(3): 274–290. http://www.jstor.org/stable/190357.

Meinzen-Dick, Ruth Susella, Nancy Johnson, Agnes R. Quisumbing, Jemimah Njuki, Julia A. Behrman, Deborah Rubin, Amber Peterman, and Elizabeth Waithanji. 2011. *Gender, assets, and agricultural development programs: A conceptual framework.* CAPRi working paper no. 99. International Food Policy Research Institute: Washington, DC.

OECD DAC Network on Gender Equality. 2018. *Aid to gender equality and women's empowerment an overview.* https://www.oecd.org/dac/gender-development/Aid-to-gender-overview-2018.pdf

———. 2020. *Aid Focussed on Gender Equality and Women's Empowerment: A snapshot of current funding and trends over time in support of the implementation of the Beijing Declaration and Platform for Action.* https://www.oecd.org/development/gender-development/Aid-Focussed-on-Gender-Equality-and-Women-s-Empowerment-2020.pdf

Pyburn, Rhiannon, and Anouka Van Eerdewijk. 2021. Advancing gender equality through agricultural and environmental research: Past, present, and future. https://ebrary.ifpri.org/utils/getfile/collection/p15738coll2/id/134727/filename/134936.pdf.

Quisumbing Agnes R., Ruth Susella Meinzen-Dick, Terri I. Raney, Croppenstedt André, Julia A. Behrman, and Amber Peterman. 2014. Closing the knowledge gap on gender in agriculture. In *Gender in agriculture: closing the knowledge gap*, ed. Quisumbing Agnes R., Ruth Susella Meinzen-Dick, Terri I. Raney, Croppenstedt André, Julia A. Behrman, and Amber Peterman. Washington, DC: The Food and Agriculture Organization of the United Nations and International Food Policy Research Institute.

Quisumbing, Agnes R., Lauren Pandolfelli. 2010. "Promising approaches to address the needs of poor female farmers: Resources, constraints, and interventions." *World Development,* 38 (4): 581–592.

Women in Climate Action Group. 2021. *Women in finance climate action group report.* https://www.aviva.com/newsroom/news-releases/2021/11/cop26-women-excluded-in-fight-against-climate-change/

CHAPTER 6

The Impact of Gender-Responsive Energy Policies on Macroeconomic Outcomes

Natalia Pastori Curbelo and Kaleab Kebede Haile

6.1 Introduction

In recent decades, there has been growing interest among scholars and practitioners in the relationship between energy consumption and economic growth, as well as the extent to which energy policies are gender-responsive. Historically, the discovery of coal, oil, and gas in the eighteenth century in Europe dramatically transformed industrial production, creating vast new sources and reserves of energy and spurring unprecedented economic activity and advancement. Since the Industrial Revolution, the strength of a country's energy sector and its economic growth have been intertwined and today these interconnections are a central feature of geopolitical discourse. A comprehensive literature review reveals considerable research and evidence confirming the relationship between energy consumption and economic growth (Zahid 2008; Yilanci et al. 2021). Much of that research also highlights the importance of crafting

N. P. Curbelo (✉) • K. K. Haile
Includovate, Queanbeyan East, NSW, Australia
e-mail: natalia.curbelo@includovate.com; kaleab.haile@includovate.com

K. Drucza et al. (eds.), *Centering Gender in the Era of Digital and Green Transition*,
https://doi.org/10.1007/978-3-031-38211-6_6

and implementing appropriate policies to support and meet the needs of energy consumers in order to catalyze economic growth. For instance, Topolewski (2021) assessed the correlation between energy consumption and economic growth in 34 European countries between 2008 and 2019, concluding that a short-term increase in energy production will result in a statistically significant increase in energy consumption. In the long-term, this impact is also statistically significant and positive. Since increases in production tend to lead to increases in energy consumption, Topolewski (2021) stresses the importance of countries considering the impact energy policies can have on people's lives and on sustainable development. These findings also confirm results from an earlier study by Zahid (2008), which investigated the causal relationship between gross domestic product (GDP) and different types of energy consumption in Pakistan, India, Sri Lanka, Bangladesh, and Nepal. Zahid established a unidirectional positive Granger causality which ran from the consumption of various energy sources (coal, petroleum, etc.) to GDP. These results also find common ground with Esen and Bayrak (2017) who examined the effects of energy consumption on economic growth by means of a panel data analysis of 75 net energy-importing countries between 1990 and 2012. The results of both the panel data and the individual country analyses indicate that there is a positive and statistically significant relationship between energy consumption and economic growth over the long-term. In short, as national import-dependency decreases, energy consumption contributes more to economic growth. Research is limited with regard to the role of gender-responsive energy policies in economic growth, even though this is a subject of increasing importance and interest in international development discourses. However, what seems evident is that gender-responsive energy policies have the potential to foster female entrepreneurship, employment, and ultimately economic growth if progressive steps are taken to make links between energy and gender equality (Wilhite 2017). Indeed, since the Fourth World Conference on Women, held in Beijing in 1995, gender mainstreaming has become both a goal and a methodology for achieving gender equality in development processes (Clancy and Mohlakoana 2020). Nonetheless, there are fears that gender mainstreaming efforts have stalled, at least in terms of addressing the gender equalities within defined economic sectors such as energy—though recently there have been indications of more positive change (ibid). A detailed examination of the intersection between energy and gender issues in East and Southern Africa by Canadian Pacific Consulting Services (CPCS) Transcom International

Limited (2021) concluded that energy tends to be addressed through a technical or economic lens, with little attention to its social dimensions. This means that energy policies in East and Southern Africa are often gender-blind, failing to consider how gendered roles lead to different energy needs. This blindness helps reinforce gender discrimination in regard to energy access and related economic opportunities (ibid).

A seminal work by Morris et al. (2019) is also insightful in this regard. Their research drew on insights generated from detailed assessments of energy and gender across the Economic Community of West African States (ECOWAS) and revealed that gender equality and energy access are closely interconnected issues, which—if properly addressed together—can support sustainable economic development. They also established that policies which consider gender roles, differences, preferences, and socio-cultural norms can help maximize the impact of energy projects, programs, and initiatives. Morris et al. (2019) further identified four primary drivers for creating a more gender-inclusive policy environment. These include viewing access to energy as a basic human right that should be available to all; understanding that gender mainstreaming is necessary for the success of energy policies, programs, and initiatives; appreciating that markets and investments can operate more effectively and profitably only if women are participating at all levels in the energy value chain (as business owners, employees, and customers); and understanding that women can help tackle the global energy crisis if fully engaged as producers, suppliers, and decision-makers in the energy sector (ibid).

This research is also supported by a 2020 United Nations Environment Programme (UNEP) report, which asserted that the energy needs, priorities, and capacities of women, men, girls, and boys should be integrated into national energy policies and strategies to make sure efforts reflect realities on the ground. In addition, understanding the underlying social behaviors and roles of men and women in decision-making can support the successful implementation of renewable energy programs and policies (UNEP 2020). This suggests that gender-responsive energy policies and strategies should be viewed as a potential lever for addressing gender inequalities across the energy value chain and can have positive collateral impacts on environmental sustainability, social inclusivity, and economic growth (as also suggested by Morris et al. 2019).

In this chapter, we contend that gender mainstreaming should be a core part of policy and practice in the energy sector and included as a matter of course in national planning and programming for development. The empowerment of both men and women is an important step toward addressing gender-related barriers that hinder the equal participation of marginalized social groups (such as women and girls) and expand access to energy in developing countries. The paucity of empirical evidence showing the positive impact of gender-responsive national energy policies on macroeconomic outcomes (such as labor force participation and economic growth) creates a need for more evidence on the role of gender-sensitive policies and on how gender roles translate into differentiated energy needs and impact women's participation in the national labor force and economic growth. Thus, we use the review of energy policies in East and Southern Africa conducted by UN Women, United Nations Development Programme (UNDP), and UNEP, to identify whether or not gender-related issues are included in policies and—if so—how they are included (i.e. under which theme, pillar, or cohort). Fifteen countries in East and Southern Africa were included in the review (Botswana, Kenya, Lesotho, Madagascar, Malawi, Mauritius, Mozambique, Namibia, Rwanda, Somaliland, South Africa, Swaziland, Tanzania, Uganda, and Zimbabwe). We also used the World Bank's World Development Indicators (WDI) database to find information on labor force participation and economic growth variables. Swaziland is not in the WDI database and was thus excluded from our analysis. Below we therefore present our assessment of the connections between gender-responsive national energy policies and macroeconomic outcomes across 14 East and Southern African countries.

6.2 The Impact of Gender-Responsive Energy Policies on Women's Employment

As shown in Figs. 6.1a–c, there are different correlations between gender-responsive national energy policies (as proxied by the number of references to gender-related issues) and the employment of women in three different sectors. In agriculture, the number of references to gender is negatively correlated with women's employment in the sector, while in the industry and service sectors it is positively correlated. To put it another way, gender references are associated with decreases in female employment in the agriculture sector and increases in the industry and service sectors.

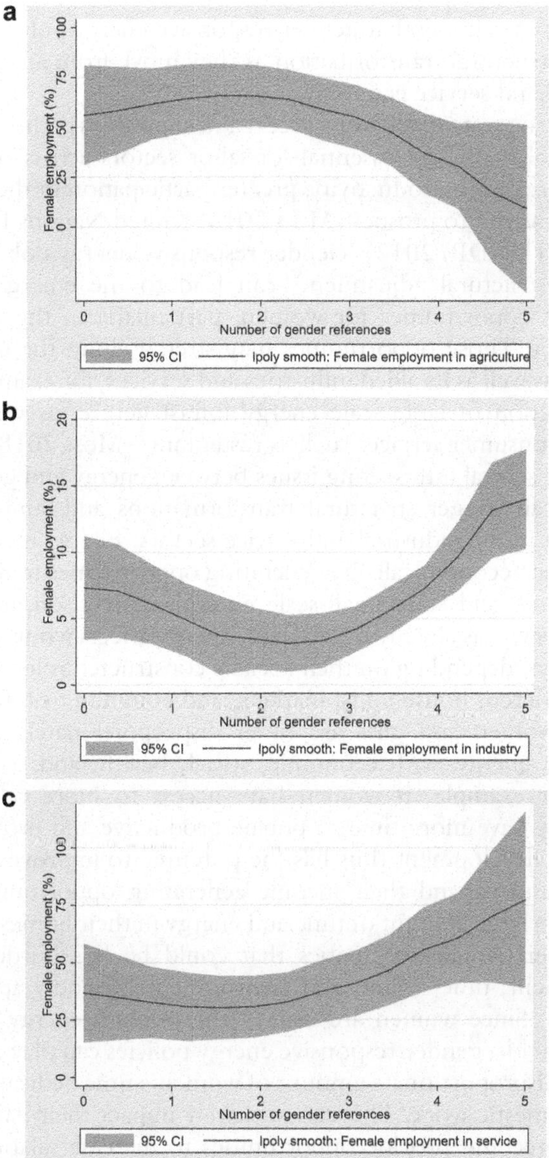

Fig. 6.1 (**a**) Number of references to gender in national energy policies and female employment in the agriculture sector. (**b**) Number of references to gender in national energy policies and female employment in the industry sector. (**c**) Number of references to gender in national energy policies and female employment in the service sector

Figures 6.1a–c imply that gender-responsive energy policies may play a role in the structural transformation as they move from an agriculture to an industrial and service economy.

Research suggests that energy access and supply—whether from renewables or fossil fuels—are essential for labor sectors across the economy, allowing increases in productivity, greater participation in the labor force, and new industries to prosper (Moss 2018; United Nations Development Programme (UNDP) 2012). Gender-responsive energy policies that promote such structural adjustments can lead to the emergence of new employment opportunities for women, particularly in the industry and service sector. Those opportunities may stretch along the energy sector value chain as well as its allied industries and services, for example, through the construction of large-scale energy infrastructure, as well as health, social, and consumer services such as restaurants (Moss 2018).

There are several intersecting issues between energy and gender that, if addressed, can trigger structural transformations and stimulate greater employment in the industrial and service sectors. Energy investments can benefit women economically by generating opportunities to develop skills, access training, and start small-scale enterprises (Nelson and Kuriakose 2017). However, as already mentioned, energy affects women and men in different ways, depending on their socially constructed roles and responsibilities in different households, markets, and communities. Research suggests that women and girls are often disproportionately impacted by shortages of energy, with economic, social, health, and security consequences; for example, if women have access to more reliable energy sources, they have more time to pursue productive and profitable activities. Energy development thus has the potential to improve the situation of women and expand their income-generating opportunities. Women often spend a large amount of time and energy in their homes fulfilling the needs of their families—activities that could be less burdensome with energy-efficient, time-saving, and environmental-friendly appliances and mechanisms. Since women are usually the primary energy managers in their households, gender-responsive energy policies can play a part in alleviating the disproportionate amount of women's time dedicated to unpaid care and domestic work. This can positively impact their participation in the labor force and provide opportunities to increase salaried work and engage in female-driven microenterprises (World Bank and UN Women 2018; Wilhite 2017; Nelson and Kuriakose 2017; Majula 2019).

Furthermore, freeing up women's time and energy can have positive impacts on their education, social capital, and well-being, allowing them to participate more fully in educational activities and build social capital through community interactions (Nelson and Kuriakose 2017). Better access to energy is associated with improved literacy in women (Wilhite 2017), which in turn can open up opportunities, employment, and income in less traditional economic sectors, such as the service sector. (The latter is a key source of employment across Africa but tends to select workers with a higher than average level of education—see Baccini et al. 2022.) In addition to providing new economic and educational opportunities for women, gender-responsive energy policies can help foster women's participation in the renewable energy sector. Here, they can influence regulatory labor conditions, including those designed to facilitate women's and men's participation in this emerging sector. Their participation also, of course, enhances gender diversity in the energy sector as a whole (Prebble and Rojas 2017).

6.3 The Impact of Gender-Responsive Energy Policies on Economic Growth

Figures 6.2a–c show that gender-responsive national policies are positively correlated with economic growth (as measured by GDP per capita). These findings suggest that gender-responsive energy policies may have a positive impact on economic growth in East and Southern Africa, perhaps by initiating structural adjustments that promote more employment, particularly of women in the industry and service sectors. Figures 6.2a–c show that countries that consider gender within their national energy frameworks may stimulate productivity, growth, and a higher GDP.

Improved access to energy can be a catalyst for economic development (measured through GDP) in East and Southern African countries due to its potential to increase the labor force. Our findings show that gender-responsive policies can play a part in freeing up women's time, increasing their access to education and participation in profitable activities, as well as boosting communities' social capital. Such elements can contribute to an increase in employment rates, which—in turn—can help grow the economy and GDP.

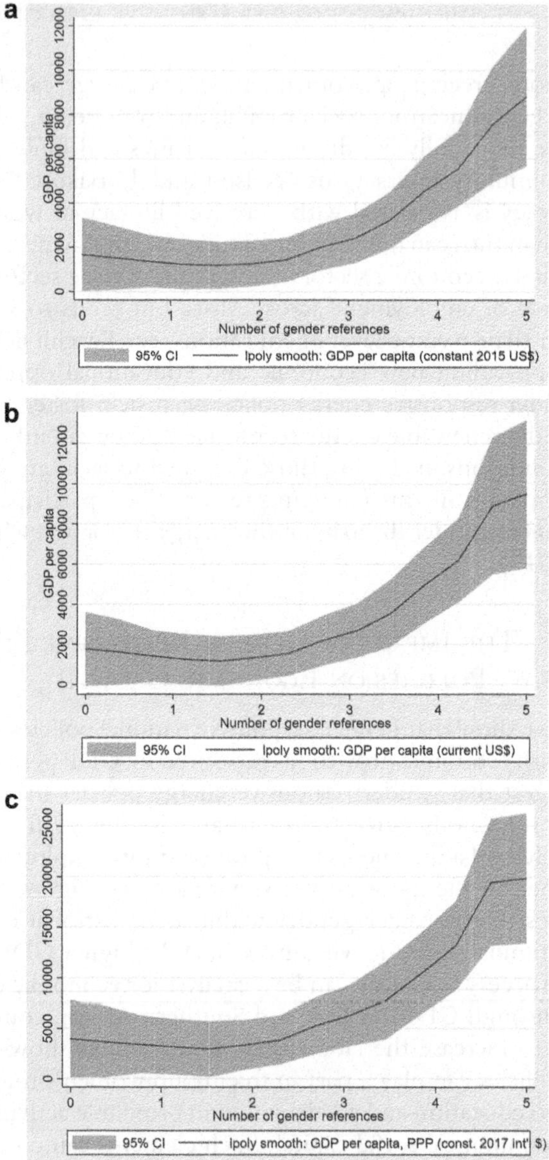

Fig. 6.2 (**a**) Number of references to gender in national energy policies and GDP per capita (constant 2015 US$). (**b**) Number of references to gender in national energy policies and GDP per capita (current US$). (**c**) Number of references to gender in national energy policies and GDP per capita (constant 2017 US$)

6.4 The Impact of the COVID-19 Pandemic

The COVID-19 pandemic put a tremendous strain on the global economy, leading to health, economic, and social crises in many parts of the world and huge impacts on labor markets. It also impacted the energy sector and stalled many LMICs' efforts to ensure affordable, reliable, sustainable, and modern energy. The pandemic led to a shift in priorities and—in many countries—to a decrease in the financial resources being directed at the energy sector. Furthermore, as Cozzi et al. (2020) highlight, millions of households in rural or peri-urban areas slipped back into using traditional fuels such as charcoal, kerosene, or fuelwood, with their associated negative impacts on health, women, and socio-economic development.

Studies developed by the International Network on Gender and Sustainable Energy (ENERGIA 2020) and the United Nations Economic Commission for Europe (UNECE 2022) highlight the impact of the pandemic on women's social and economic inequalities, including threats to those working in the energy sector, in small business, or in the informal sector. Pandemic measures, such as lockdowns and social distancing, school closures, and homeschooling, increased and intensified the need for unpaid care work within households. These burdens largely fell on women, creating particularly heavy weights for women without access to energy, electrical appliances, and other time and labor-saving devices. This in turn may have impacted women's ability to cope with the COVID-19 crisis and hindered their continuation (or initiation) of home-based businesses or remote work (ENERGIA 2020). Enabling the availability and adoption of such devices, as well as access to energy, will be key to building future resiliency in many LMICs. Gender-responsive energy policies may play a role in protecting women from future crises and external shocks and developing mechanisms for adaptability and resilience. Further analysis is still required, however, to understand their role in the context of the COVID-19 pandemic.

6.5 Concluding Remarks and Recommendations

In this chapter, we empirically examined the association between gender-responsive national energy policies and macroeconomic outcomes in 14 East and Southern Africa countries, measuring the structural transformations in the labor force across key economic sectors as well as economic growth (measured by GDP per capita). Our research demonstrated that gender-responsive policies are positively correlated with the proportion of women employed in the industry and service sectors, but negatively correlated with the proportion of women employed in the agriculture sector. These findings highlight that gender-responsive policies are positively correlated with increased employment in general and, in particular, to more women participating in the agriculture, industry, and service sectors. Our findings show a positive association between gender-responsive national policies and economies structural transformation. Likewise, they suggest that gender-responsive national energy policies positively impact macroeconomic outcomes in many East and Southern African countries. The positive association between gender-responsive policies and economic growth could be due to their influence on women's employment in the industry and service sectors (rather than in the agriculture sector).

However, this analysis cannot assure causation. There may be unobserved differences between the sample countries that play a role in the associations observed. For instance, more frequent references to gender in national energy policies may be due to efforts by donors or international development partners to stimulate more inclusive policies, infrastructure investments, and desirable macroeconomic outcomes. Further studies are needed, therefore, to examine and establish a causal relationship between gender-responsive national energy policies and macroeconomic outcomes and economic growth. These would require controlling for country-level characteristics and the effects of economic growth on the gender-responsiveness of national energy policies.

References

Baccini, Leonardo, et al. 2022. Services, jobs, and economic development in Africa. *VoxEu Column. CEPR* 38: 147.

Clancy, S. Joy, and Nthabiseng Mohlakoana. 2020. Gender audits: An approach to engendering energy policy in Nepal, Kenya and Senegal. *Energy Research & Social Science* 62: 101378. https://doi.org/10.1016/j.erss.2019.101378.

Cozzi, L., et al. 2020. *The Covid-19 crisis is reversing progress on energy access in Africa*. Paris: IEA.

CPCS Transcom International Limited. 2021. Report on implementing harmonized regulatory/technical frameworks and synthesized renewable and energy efficiency strategies in the EA-SA-IO Region: Energy Policy Gender Assessment.

ENERGIA. 2020. Gender and energy at Centre stage in COVID-19 battle: Powering a more gender-equal recovery.

Esen, Omer, and Meting Bayrak. 2017. Does more energy consumption support economic growth in net energy-importing countries? *Journal of Economics, Finance and Administrative Science* 22 (42): 75–98.

Majula, M. 2019. How gender-sensitive are India's energy policies? *Commentary* 3: 18–21.

Morris, Ellen, Jennye Green, and Victoria Healey. 2019. *Blueprint guide for creating gender-sensitive energy policies*. Lisbon: Clean Energy Solutions Center and National Renewable Energy Laboratory.

Moss, Todd. 2018. *Job creation and energy in Africa*. Energy for Growth Hub: Memo.

Nelson, Sibyl, and Anna Kuriakose. 2017. *Gender and renewable energy: Entry points for Women's livelihoods and employment*. Climate Investment Funds.

Prebble, Maria, and Ana Rojas. 2017. *Energising equality: The importance of integrating gender equality principles in national energy policies and frameworks*. Washington: IUCN Global Gender Office.

Topolewski, Łukasz. 2021. Relationship between energy consumption and economic growth in European countries: evidence from dynamic panel data analysis. *Energies* 14 (3565): 1–12. https://doi.org/10.3390/en14123565.

United Nations Development Programme (UNDP). 2012. *Integrating energy access and employment creation to accelerate progress on the MDGs in Sub-Saharan Africa*. New York: UNDP.

United Nations Economic Commission for Europe (UNECE). 2022. *Energy transition and post-Covid-19 socioeconomic recovery: Role of women and impact on them*. ECE ENERGY SERIES no. 145. Geneva: UNECE.

United Nations Environment Programme (UNEP). 2020. *Gender integration in renewable energy policy: A guideline for renewable energy policy and decision makers*. New York: UNEP.

Wilhite, Harold. 2017. Gender implications of energy use and energy access. EEG state-of-knowledge paper Series. Paper No: Gender Theme.

World Bank —Energy Sector Management Assistance Program (ESMAP) and UN Women. 2018. *Policy brief 12. Global Progress of SDG 7 - energy and gender*. New York: UNDP.

Zahid, Asghar. 2008. Energy–GDP relationship: A causal analysis for the five countries of South Asia. *Applied Econometrics and International Development* 8 (1): 167–180.

CHAPTER 7

Women in the Transport Sector: From Mobility to the Profession

Jasmina Bunevska Talevska, Amira Kaddour, and Alina F. Burlacu

7.1 Introduction

Women and men have different travel requirements and different transport options due to their different roles in society. They face distinct travel-related challenges, as well as stereotypes and prejudices about their

J. B. Talevska
University "St. Kliment Ohridski", Bitola, Republic of Macedonia
e-mail: jasmina.bunevska@tfb.uklo.edu.mk

A. Kaddour (✉)
National School of Advances Sciences and Technologies, University of Carthage, Amilcar, Tunisia

Includovate, Tunis, Tunisia
e-mail: amira.kaddour@enstab.ucar.tn

A. F. Burlacu
Global Road Safety Facility, Washington, DC, USA
e-mail: fburlacu@worldbank.org

K. Drucza et al. (eds.), *Centering Gender in the Era of Digital and Green Transition*,
https://doi.org/10.1007/978-3-031-38211-6_7

abilities to participate in the transport sector. Furthermore, societal expectations (e.g. that women should oversee household chores and raise children) limit the amount of time women are able to spend traveling. The 1948 Universal Declaration of Human Rights sets out (in Article 13) the right to freedom of movement.[1] Without full freedom of movement, men and women cannot equally access socioeconomic opportunities, employment, services, or cultural resources. Without full freedom of movement, therefore, gender equality will not be achieved, and progress will be limited toward several Sustainable Development Goals (SDG) including SDG 5. A gendered analysis of mobility looks at the subject from three key angles: first, considering, if women have the right to travel; second, considering women's professional participation in the transport sector; and third, considering how road safety issues may differently affect women (women are more affected by road accidents but contribute less to them).

7.2 Mobility as a Human Rights Issue

Human mobility boosts economic growth, reduces inequalities, and brings disparate societies together. A study by the International Monetary Fund and the World Bank, for instance, shows that an increase in the percentage of migrants in the population of high-income countries boosts the income per capita by 2 percent. If immigrants augmented the labor force of rich countries by 3 percent, global GDP would increase by US$356 billion by 2025 (Ullah et al. 2021). To meet their daily needs (i.e. for work, leisure, culture, education, health, and participation in public life), human beings need transport options that are readily available, as well as reasonably safe, reasonably priced, and reasonably quick. According to travel behavior theory, this movement is a decision-making and optimization process that aims to maximize the utility of all these elements. However, despite progress in human rights, women's mobility is still limited worldwide, hindering their development. Beyond the socioeconomic and developmental needs that the mobility factor satisfies toward the eradication of poverty and disparities, the Covid-19 crisis has nuanced the crucial place of mobility in the shared daily life of human beings as it serves to allow satisfying many needs.

[1] "Everyone has the right to freedom of movement and residence within the borders of each state" and "Everyone has the right to leave any country, including his own, and to return to his country."

7.3 Women's Mobility: Their Specific Needs and Concerns

The ability to move freely is a fundamental human right that supports the attainment of multiple SDG and promotes an inclusive society. In September 2015, UN Women and the People's Republic of China jointly hosted a Heads of State Meeting on Gender Equality and Women's Empowerment at the United Nations headquarters in New York. The goal of the meeting was to secure tangible commitments and prioritize gender equality, women's rights, and empowerment in the new global agenda.

Despite such international engagement, mobility policies and means of transport are still unequal in many countries. Men and women travel in different ways and for different reasons (Blumenberg et al. 2022), use and acquire vehicles differently, and participate differently the transport sector as workers and managers (ILO 2013). Their contributions to planning and executing transportation infrastructure, services, and vehicles also vary. The following analysis will highlight these variations.

The mobility of women is crucial for a country's economic growth. When women have access to transportation, they can travel to work and access other income-generating opportunities beyond their locality. This increases their chance for employment and income. However, traditional transportation methods, such as walking or biking, are often limited for women by cultural, social, and economic factors. In some cultures, women may not be allowed to travel alone, or they may not have access to the same transportation options as men. Additionally, women may not have the same level of education or income as men, which can further limit their mobility. New transportation methods, such as ride-or bicycle-sharing services or programs, can provide more options and greater accessibility for women. For example, ride-sharing services can provide women with a safe and reliable mode of transport while bike-sharing programs can give women the opportunity to travel short distances quickly and easily. However, these new transportation methods may also have their own limitations (services may not be available in certain areas, for example, or may be too expensive for some women to afford, or not acceptable in certain cultures). In 2007, a few European Union (EU) member countries introduced a common bicycle system, which gained popularity in 2016 with the addition of new services for cars, motorcycles, bicycles, and scooters. As these services are relatively recent, data on their usage is limited, but research does show significant differences in usage between men and women, with educated and upper-middle-class men most likely to benefit from such services. With more alternative mobility solutions available that better meet women's needs for shorter and faster journeys, questions arise about why women do not benefit as much from shared mobility and if it can be an alternative to traditional transportation. The Transport Innovation Gender Observatory (TinnGO) is an EU-funded project that aims to develop gender-sensitive and diversity-sensitive smart mobility solutions for sustainable European transport. TinnGO studies transportation as a means of achieving gender equality, analyzes traffic patterns, and applies an intersectional analysis that considers sociocultural and gender dimensions. The project aims to use diverse transport data and assessment tools, to model and plan new mobility policies and services in order to create a path toward gender-sensitive smart mobility in European transport. TinnGO serves as a starting point for innovations around transport and gender. In TinnGO (2017) conducted interviews that shed light on the distinct behaviors and challenges faced by women when utilizing shared transportation. Most significantly, women reported experiencing physical and verbal attacks, and frequently felt insecure, especially at certain times of day and for certain travel purposes. Fear and the physical reality of attacks have a substantial impact on women's transportation decisions, including their clothing choices. Data from the Paris Institute indicates that 51.4 percent of women were fearful when using public transportation (compared to 23.3 percent of men), reflecting an upward trend over the past decade. This finding is consistent with surveys by Uber and Harris Interactive, which showed that 52 percent of women prefer using hired cars to return home rather than public transportation, with a mere 11 percent opting for the latter. Although many women regard digital platforms and solution like Uber as relatively safe, women have frequently been the victims of violence through them, as shown in numerous investigations. For example, in France in November 2019, the #UberCestOver Twitter campaign revealed stories of users who had been harassed and abused while using public transportation. This campaign prompted Uber to improve their app's functionality to ensure passenger safety. A vital aspect of assessing transportation is passenger safety. The results of the TinnGO project demonstrate that when women are deciding which type of public transport to use, their decisions are heavily influenced by safety issues. Prioritizing safety (including when designing paths, stations, and streets) is thus likely to increase the number of public transport users.

Is Transport Equally Available to Women?

Transport gender disparities become especially apparent when considering private cars. While there is no noticeable gender difference in the availability of private cars, less women own cars and hold driver's licenses, therefore private cars are less accessible for women. These disparities are particularly pronounced in rural areas and among poorer households. Similarly, women are less likely to own or use bicycles. Women rely on public transport more frequently than men and are more likely to be impacted when public transport is limited, as it is in rural and less developed areas, where gender mobility inequalities become more evident (Blumenberg et al. 2022).

How and When Women Miss Out

Women are less likely to hold driver's licenses than men, compared to industrialized and developed EU countries and, as a result, are more reliant on lifts, public transportation, and walking. They are thus less likely to be fully mobile, and this impacts their access to a variety of services, including healthcare, education, cultural, and sporting activities. Women are particularly likely to miss out on the latter two when transport is difficult or inaccessible) and on healthcare and education if they live more than three kilometers from services.

Which Types of Transport Do Women Prefer?

Both men and women find walking or taking a car (whether as a driver or passenger) the most comfortable and attractive modes of transport, although the latter is statistically the most dangerous mode. Real-time information about delays to travel are a key factor in whether transportation modes are considered attractive or not, regardless of gender, age, or education level. Women face higher safety hazards on all modes of transportation, but public transport causes increased risks of pickpocketing, harassment, and discrimination. Both men and women attribute their discomfort with public transportation to poor infrastructure and service. Men are more likely to move around to engage in social activities—going to restaurants, cafes, meeting friends, and so on—while women are more likely to go shopping or attend cultural events. Other factors, including social status and education, influence the frequency with which transportation is used.

Women's Place in the Transport Sector

Infrastructure and facilities have a considerable impact on the way people move, particularly women. Involving more women in transportation planning can therefore help improve their access to key services (medical, educational, administrative, etc.).

Employment in the Transport Sector

According to the 2020 OECD report, women are far less likely than men to be employed in the transport sector and account for just 17 percent of employees. Women are more commonly employed by state-owned corporations, while men tend to work in private businesses, and women's engagement as entrepreneurs in this sector is notably low. When comparing salaries, women generally receive higher gross salaries than men, but men have higher net salaries. Both lack social welfare rights, such as pensions and healthcare insurance (especially drivers and workers). Several key factors contribute to these gender inequities in the transport sector, including biased employment policies that limit women to administrative roles; prejudices and stereotypes that favor men as drivers; failures to implement gender equality laws; and the absence of specialized schools for women in the sector.

Within the transport sector, many businesses do not actively address gender differences or prioritize gender equality. Ensuring women's safety at work, particularly in client services, is a widespread concern, and often hindered by a lack of protocols, backup support, and safety equipment. There is insufficient data to consider salary differences between men and women, their career development, or their access to resources or gender rights in more detail, and an associated lack of information to assess whether transportation services meet women's specific needs. It is vital to protect women from gender-based discrimination, but recognition of their rights, and initiatives to fulfill those rights, is often lacking. In addition, infrastructure sometimes fails to accommodate women's demands and is often inaccessible for persons with disabilities. In the field of air transport, women are well represented. This is a contrast to other modes of transport, where cultural and political factors often discourage women from pursuing careers in the sector. Stereotypes play a significant role in perpetuating imbalances and creating negative perceptions. These stereotypes suggest that women lack the physical skills to handle stress, the resilience needed for long journeys, and/or the ability to handle mechanical breakdowns, which drivers and workers often face the transportation industry. In addition, there is a misconception that women are unable to travel at night and handle the potential dangers associated with being on public streets. These stereotypes contribute to the underrepresentation of women in the transport sector and hinder their professional opportunities.

7.4 Analyzing the Place of Women in the Transport Sector: The Case of European Countries Toward Inclusion

Europe offers some exemplary models of how the transport sector can modernize and improve road safety. Robust road safety measures in Europe include intelligent transportation systems, well-maintained road networks, stringent traffic laws, advanced infrastructure, and vehicle safety features, and strict enforcement mechanisms. The Sweden model of zero accidents is considered a worldwide benchmark. European countries have also increased their efforts to plan and develop comprehensive transport systems more inclusively. That has included involving diverse stakeholders in decision-making processes to ensure inclusive and equitable transport systems that consider the diverse needs of different user groups. This collaborative approach has led to the creation of safer, more accessible, and more inclusive transportation options for all, including women, children, the elderly, and people with disabilities. In November 2022, the European Commission passed a groundbreaking regulation, mandating that the Boards of European companies should increase women's representation to 40 percent by 2026. Women are underrepresented in leadership roles across diverse fields, including the transport industry, and enhancing female representation here could have broader implications for achieving equality across industries. Numerous studies demonstrate that women and men have distinct mobility needs, yet current transportation systems primarily cater to male requirements (WEF 2023). Despite these important efforts, gender inequality in the transport sector is important; this can be explained by a dearth in previous research that assess the gender lens in the transport sector and the requirements that the female population has when it comes to travel and transport. The gendered issues of transportation have become more prominent in recent years, particularly in France,

Sweden, and Scotland, as increasing numbers of professional women are employed and belong to associations within the transport industry (International Transport Forum 2011). Numerous programs and projects have tried to improve women's access to services and modes of transportation, as well as to better involve women in transport-related activities. The International Transport Forum has put forward several recommendations to improve the situation of women. Firstly, it suggests considering gender-specific statistics in relation to key transport aspects (such as access to different modes of travel, costs, quality of services, and travel characteristics). Second, it emphasizes the importance of involving women in transport planning and decision-making. Third, it supports increasing female employment in the transportation sector to foster gender diversity and inclusivity. "Inclusion" in the mobility and transport sector entails making transportation infrastructure and services accessible and beneficial to people of all ages, genders, abilities, and socioeconomic backgrounds. That requires that we recognize, address, and meet the needs of specific marginalized groups, such as persons with disabilities, the elderly, low-income communities, and women. Inclusive transportation policies can reduce barriers, improve connectivity, expand mobility options, improve physical accessibility, affordability, safety, and convenience, and provide equal opportunities to all (in education, employment, healthcare, and recreational activities). Analyzing the current situation, two problems become apparent in terms of inclusive mobility. First, current transport systems and services tend to cater for the needs of "normal" people, going to work in the "normal" way. That bias often fails to meet the needs of distinct groups of employees. Second, the transport sector is dominated by men (women only make up 22 percent of transport workers in Europe, e.g. see Eurostat (2021)). As mentioned above, the TinnGO project has set a starting point for gender innovations around transport and will create a path for others to follow in terms of creating gender-sensitive smart mobility. In Europe that has involved applying intersectional analyses, diversifying transport data and assessment tools, and modeling new mobility policies, planning, and services. The EU's Strategic Travel Research and Innovation Agenda (STRIA) has acknowledged the mobility differences between women and men. This observation is supported by statistics from Paris, France. Men and women there exhibit distinct patterns in travel behaviors. For example, while men tend to rely more on private vehicles (cars, motorcycles, bicycles, etc.), women are more likely to use public transport. Women also more frequently travel to work out of peak hours;

their destinations tend to be shorter, and they undertake more trips than men. The Transport Research and Innovation Monitoring and Information System (TRIMIS) is the analytical support tool for the establishment and implementation of Strategic Transport Road safety that considers a gender approach. In summary, Parisian women favor different travel modes, travel for different purposes, and their trips differ from men's both in frequency and duration.

7.5 Analyzing Road Safety Statistics and the UN Agenda for 2030: Gender Caution

According to the World Health Organization (WHO), the cost of road safety in low- and middle-income countries is equal to 2 to 5 percent of GDP, and the accident rate is 26 per 100,000 inhabitants. Despite significant declines in road fatalities in the European Union and countries such as Brazil and the Russian Federation, road fatalities remain unacceptably high, with 1.35 million people killed each year and up to 50 million injured.

Partners have been working to ameliorate these alarming statistics. For example, World Health Organization's Safe System approach, now adopted by most countries, helps analyze and monitor road safety indicators. It represents a holistic vision that analyzes the role of the components influencing the quality of mobility but also the interaction between them and the synergistic factors that exist. Women's mobility issues are also increasingly being considered in national targets and programs, as well as in the collection and monitoring of road safety data. However, rarely do these initiatives consider the specific needs of women. For example, WHO's safe system approach did not incorporate intersectional or gendered considerations. The UN's Second Decade of Action for Road Safety 2021–2030 provides an opportunity to build on the successes and lessons of previous years to save lives and facilitate the achievement of the SDGs. Twelve Road Safety Targets for 2030 have been fixed and approved by the UN—see Table 7.1. However, there is an urgent need to bolster the targets and road safety issues through gender-sensitive assessments (see Table 7.1, right-hand column).

In February 2020, a major ministerial conference on road safety was held in Stockholm, Sweden, with the participation of 140 countries and 1700 participants. It aimed to escalate progress toward the 12 objectives

Table 7.1 Road Safety Targets for 2030

Horizon 2030	*Gender Equality and Social Inclusion Assessment*
Target 2: By 2030, all countries use one or more of the UN's core road safety-related legal instruments.	Five assessments: Traffic rules, road signs, vehicle regulations, transport of dangerous goods, professional driver fatigue. **Assessment:** Gender was not considered in assessing professional driver fatigue.
Target 3: By 2030, all new roads achieve technical standards for all road users that consider road safety or meet a three-star rating or better. **Target 4:** By 2030, more than 75% of travel is (a) on roads that meet technical standards and (b) considers road safety for all road users.	To monitor the safety and quality of the road network and to draw attention to the need for improvements. Such programs have predictive ratings which look at the protective quality of various elements of a road network as well as retrospective elements which involve risk mapping and performance tracking according to specific protocols. **Assessment:** Youth, PWDs, and women are the most vulnerable
Target 5: By 2030, 100% of new[a] and used vehicles meet high-quality safety standards.[b]	No technical assessment of safety standards or specific tools for pregnant women. **Assessment:** Neutral
Target 6: By 2030, halve the vehicles traveling over the speed limit and achieve a reduction in speed-related injuries and fatalities.	**Assessment:** Youth, PWDs, and women are the most vulnerable.
Target 7: By 2030, increase the proportion of motorcycle riders correctly using standard helmets to close to 100%.	WHO's global best practice standards for motorcycle safety **Assessment:** Youth, PWDs, and women are the most vulnerable.
Target 8: By 2030, increase the proportion of motor vehicle occupants using safety belts or standard child restraint systems to close to 100%.	The standard for child seat
Target 9: By 2030, halve the number of road traffic injuries and fatalities related to drivers using alcohol and/or achieve a reduction in those related to other psychoactive substances.	Level 1 (less than 0.05 g) Level 2 (0.05 g or more, but less than 0.08 g) Level 3 (0.08 g or more, but less than 0.15 g) Level 4 (0.15 g or more) **Assessment:** Youth, PWDs, and women are the most vulnerable

(*continued*)

Table 7.1 (continued)

Horizon 2030	*Gender Equality and Social Inclusion Assessment*
Target 10: By 2030, all countries have national laws to restrict or prohibit the use of mobile phones while driving.	**Assessment:** Neutral
Target 11: By 2030, all countries enact regulations for professional drivers (driving time, rest periods, etc.) and/or accede to international/regional regulations in this area.	**UN recommendations**: (1) Modernize the training by prioritizing road safety (including protecting vulnerable road users and using driver assistance systems; optimizing fuel consumption; and introducing new technology, such as e-learning options). (2) Clarify the rules on minimum ages (set out in the professional Drivers Directive/EU rules on driving licenses). (3) Ensure that training taken in a different EU country (i.e. other than the one where the person lives) is recognized in their country of residence; make the directive clearer and easier to understand. **Assessment:** No specific mention to women needs.
Target 12: By 2030, all countries establish and achieve national targets in order to minimize the time interval between road traffic crashes and the provision of first professional emergency care.	The UN General Assembly declared the years 2011–2020 as a Decade of Action for Road Safety. Member states also included two specific targets on road safety in the SDGs: • SDG target 3.6. Seeks to reduce road traffic deaths and injuries by 50% by 2020 • SDG target 11.2. Aims to provide access to safe, affordable, accessible, and sustainable transport by 2030. **Assessment:** Youth, PWDs, and women are the most vulnerable

[a]Defined as produced, sold, or imported

[b]Such as the priority UN Regulations, Global Technical Regulations, or equivalent recognized national performance requirements

and reduce the rate of road accidents worldwide by 50 percent. To do so, we will need to shift toward safe, sustainable, and intelligent cities, based on a consensus of good human reflexes to be adopted and a multidimensional governance framework.

Considering Women in the 12 Targets According to a Cordellieri et al. (2016) women have less risky driving habits, tending to drive more carefully, more slowly (speed is the main cause of road accidents), and adhere more to traffic rules. Despite this, they are more vulnerable as pedestrians to road accidents. In short, whilst women contribute less to accidents, they are more likely to be a victim of them. The 12 targets must, therefore, consider road safety in relation to women's mobility and ensure sex-disaggregated indicators are collected and available to inform analysis and policy.

7.6 Disaggregating Data: The Key to Improving and Monitoring Women's Mobility

Disaggregating data is critical to improving women's mobility and road safety (as it is to improving many other areas, sectors, and fields). Here, as elsewhere, the absence of good, disaggregated data impacts understanding, monitoring, and accountability and hinders our ability to conceptualize solutions and achieve progress through effective planning, policies, and program implementation.

To understand the importance of disaggregated data in road safety, it is important to first understand exactly what "aggregate data" is versus "disaggregated data." There is no one definitive source that speaks to these distinctions; however, the following definitions, collected from a multitude of sources, provide a simplified overview of aggregate versus disaggregated data. Aggregate data is data that has been collected from two or more sources. Disaggregated data is data that has been extrapolated or taken from aggregated data and divided into smaller data units. Disaggregating data is another critical step to gaining increased knowledge from collective or aggregated information. It involves delving more deeply into a set of results to highlight issues that pertain to individual subsets of results and/or outcomes of aggregated data.

In road safety, data that is disaggregated by gender, age, road type, road user group, urban/rural location, and administrative units helps to establish the scope of the problem, expose hidden trends, identify vulnerable road users, and make at-risk groups more visible to policymakers.

7.7 Concluding Remarks and Recommendations

Understanding and improving women's mobility involves collecting data on it. Such data can be of particular use in low and middle-income countries, where there are a high number of road accidents. However, currently there is a lack of infrastructure and limited capacity[2] to collect reliable and accurate data, which impacts not only the achievement of inclusive mobility, but many other social and economic objectives (health, education, equality). Mobility is a cross-cutting factor that impacts all areas of the economy, but it is also a factor of social well-being, which has been highlighted once again by the COVID-19 crisis. The economic inclusion of women requires the development of their skills, their level of education, the knowledge of their rights and obligations, and the improvement of their mental and physical health. Without safe, accessible, and continuously improving mobility, women will not be able to overcome the sociocultural barriers imposed in many parts of the world. From a gender equality perspective, women's access to the transport sector is still unequal today; women face persistent gender stereotypes that discourage their employment in many transport sector branches. Equality objectives must be set as part of the company's social responsibility indicators.

References

Blumenberg, Evelyn, Andrew Schouten, and Anne Brown. 2022. Who's in the driver's seat? Gender and the division of car use in auto-deficit households. *Transportation Research Part A: Policy and Practice* 162: 14–26. https://doi.org/10.1016/j.tra.2022.05.017.

Cordellieri, Pierluigi, Francesca Baralla, Fabio Ferlazzo, Roberto Sgalla, and Laura Piccardi. 2016. Gender effects in young road users on road safety attitudes, behaviors and risk perception. *Frontiers in Psychology* 7: 1412 https://doi.org/10.3389/fpsyg.2016.01412

European Commission. Women in transport. https://transport.ec.europa.eu/transport-themes/social-issues-equality-and-attractiveness-transport-sector/equality/women-transport_en#:~:text=European%20Commission%20encourages%20women%20to,of%20transport%20workers%20are%20women

Eurostat. 2021. Transport database. https://ec.europa.eu/eurostat/web/transport

ILO. 2013. Women in the transport sector. Transport policy brief. https://www.ilo.org/wcmsp5/groups/public/%2D%2D-ed_dialogue/%2D%2D-sector/documents/briefingnote/wcms_234882.pdf

[2] Road safety is often managed through national road accident registers, whose accuracy differs according to the used tools (manual collection and intelligent collection system) and the level of training and involvement of the staff in charge.

International Transport Forum Gender and Transport Discussion Paper No. 2011–11, https://www.oecd-ilibrary.org/transport/gender-and-transport_5kg9mq47w59w-en 4
OCDE. 2020. The Gender Dimension of the Transport Workforce. https://www.oecd.org/employment/the-gender-dimension-of-the-transport-workforce-0610184a-en.htm#:~:text=Women%20remain%20underrepresented%20in%20most,governments%20and%20the%20private%20sector.
Research and Innovation Agenda STRIA. https://research-and-innovation.ec.europa.eu/research-area/transport/stria_en
Transport innovation for gender observatory. 2017. https://www.tinngo.eu/about-us/initiatives-of-interest/
The Transport Research and Innovation Monitoring and Information System (TRIMIS). https://trimis.ec.europa.eu/
Transport Innovation Gender Observatory Report, H2020-EU.3.4.-SOCIETAL CHALLENGES – Smart, Green And Integrated Transport. https://transport-genderobservatory.eu/
Ullah, AKM Ahsan, Faraha Nawaz, and Diotima Chattoraj. 2021. Locked up under lockdown: The COVID-19 pandemic and the migrant population. *Social Sciences & Humanities Open* 3(1): 100126, ISSN 2590–2911. https://doi.org/10.1016/j.ssaho.2021.100126.
UN Women. The 2030 Agenda for Sustainable Development. https://www.unwomen.org/fr/what-we-do/2030-agenda-for-sustainable-development
World Economic Forum. 2023. Women in mobility: how is the sector driving gender equality? https://www.weforum.org/agenda/2023/01/women-in-mobility-sector-gender-equality/
World Health Organisation. 2019. Road safety strategy for the United Nations and its personnel: a partnership for safer journeys. https://www.who.int/road-safety/publications/UN-RS-Strategy/en/

PART III

Women Empowerment and Barriers

CHAPTER 8

Women's Empowerment and Leadership: Barriers and Perspectives

Nadia Labidi and Saida Gtifa

8.1 Introduction

Empowering women through access to education, training, and resources can lead to increased income and improved livelihoods, resulting in economic growth and development. When more women are economically empowered, it can also result in a more equitable distribution of wealth and a more stable society. Additionally, empowering women in the economy can lead to greater diversity in decision-making and problem-solving, which can improve economic outcomes. Research has shown that companies with more diverse boards and management teams tend to perform

N. Labidi (✉)
The Higher School of Communication of Tunis (Sup'Com), University of Carthage, Tunis, Tunisia
e-mail: nadia.labidi@supcom.tn

S. Gtifa
Faculty of Economic Sciences and Management of Nabeul, University of Carthage, Tunis, Tunisia
e-mail: saida.gtifa@fsegn.u-carthage.tn

K. Drucza et al. (eds.), *Centering Gender in the Era of Digital and Green Transition*,
https://doi.org/10.1007/978-3-031-38211-6_8

better financially. Women's empowerment plays a vital role in poverty reduction, as it is a key driver of development and a powerful tool to reduce poverty. When women have access to education, healthcare, and economic opportunities, they are better able to lift themselves and their families out of poverty. However, the lack of legal protection and discriminatory employment laws in some nations is still a barrier to women's participation in the workforce. Overall, 35 national economies still don't have legal guidelines to protect women in the workplace, and 104 economies have laws preventing women from working in many jobs. Despite this, women have made significant progress in recent years.

Both green and digital economies have the potential to create new opportunities for women and improve their livelihood conditions. A green economy prioritizes the use of renewable resources, sustainable practices, and environmental conservation to reduce the negative impacts of economic growth on the environment. It can create new jobs in renewable energy, sustainable agriculture, and conservation, many of which may be well-suited for women. The digital economy, also known as digital transformation, is the integration of technology into all aspects of society and the economy, including business, government, and daily life. It can create new opportunities for women in technology, e-commerce, and online services, and enables women to more easily access information and resources to start their own businesses and improve their livelihoods.

8.2 Promoting a Fair and Just Transition Through Female Entrepreneurship

Female entrepreneurs are key drivers of economic and social development, stimulating innovation and GDP growth, as well as providing direct employment. However, women are underrepresented in the business sector of most countries. For example, between 2000 and 2016, only 10% of the project initiators behind innovative companies resulting from the i-Lab competition or public research incubators were women. This percentage, which fluctuates significantly from year to year, has been on a downward trend since 2009, after a period of strong growth between 2000 and 2009. In the era of disruptive innovation and significant transformation, it is important to empower women's initiatives to increase their participation in sustainable development.

What Is Women's Empowerment?

Empowerment may be described as a "multi-dimensional social process that allows people to gain control over their own lives" (Page and Czuba 1999). It is a process that fosters power in humans, which they can use to ameliorate and transform their personal situations, their communities, and their societies, "by working on issues that they define as important" (Page and Czuba 1999). Using another definition, women's empowerment refers to "women's capacity to make strategic life choices where that ability had previously been denied them". Empowerment is important to support the positive role/place of women's and contribution at the individual, family, network, and broader levels. It involves actions that boost the status of women via literacy, schooling, training, and raising awareness (Alvarez 2013). However, this is no simple thing, and ensuring that women's skills, talents, experience, and energies are properly fostered and utilized calls for intentional moves and planned policies.

The Women's Empowerment Principles (WEPs), a joint initiative of the UN Global Compact and UN Women, were developed in 2010 to provide a holistic framework to empower women and girls. Established through global multi-partner consultations, the principles help the private sector incorporate gender equality initiatives and elements within their workplace, marketplace, networks, and communities. Enhancing business inclusiveness at every stage—from policies to operations—requires good techniques, tools, and practices. The Women's Empowerment Principles provide a "gender lens" through which businesses can survey and examine cutting-edge initiatives, benchmarks, and reporting practices. This is critical, since empowering women not only has important impacts on societies but helps develop sustainable, inclusive economies (Mangin 2014).

Female Entrepreneurship: A Key Pillar for Sustainability

Research suggests that women are more likely to prioritize environmental issues than men. Studies by Hechavarria et al. (2012) and Braun (2010) have found that female entrepreneurs are particularly concerned with green issues and often implement environmental-friendly practices in their businesses. Additionally, research by Sumathi et al. (2014) suggests that being a green business owner can provide economic and social benefits, as well as environment-orientated jobs for women. It is important to note that women are disproportionately impacted by climate change due to sociocultural and socio-political discrimination (United Nations Development Programme 2012). To address this, the SDGs promote gender equality and the empowerment of women not only as human rights issues, but as critical tools for addressing poverty and socio-environmental issues. Despite progress toward sustainable development and a green economy, however, gender inequality persists.

Gender equality and social justice are key elements in the transition to a green economy. The Organization for Economic Cooperation and Development (OECD 2008) emphasizes the need to implement policies that take gender dimensions into account in order to bridge the gender gap and promote sustainable economic growth through the full participation and utilization of women's skills and qualifications in the green economy. Despite progress in gender equality, significant gaps persist between men and women. To address this, there is a need for a policy framework that promotes equal rights and opportunities for both genders. United States Agency for International Development (USAID)'s 2016 study on gender and development have highlighted the importance of gender inclusion for development outcomes, particularly in terms of allowing women to leverage their skills and decision-making in all areas of life. In regions such as Middle East and North Africa (MENA), this means promoting the active participation of women in all aspects of governance. Webb (2015) highlights that while gender analysis and mainstreaming have led to a better understanding of gender inequality in relation to climate change, much of the research and policies in this area are based on generalized perceptions and stereotypes. Smith and Musango (2015) also argue that past strategies in the development of green economies have not adequately considered the different implications for men and women. The United Nations Environment Programme (United Nations Environment Programme 2011a, 2011b) defines the green economy as one that promotes well-being and social equity by addressing environmental risks and resource shortages. Achieving a green economy requires investments in areas that spur economic growth while also promoting resource efficiency and environmental-friendly practices. This includes considering poverty reduction and women's empowerment as crucial to sustainable development. Godwell and Mukonza argue that to effectively transition to a green economy, it is necessary to create policies and programs that promote green economy adoption for all groups. One important reason to prioritize women's empowerment and gender equality is economic development. Sustainable development cannot be achieved without the participation of women (OECD 2008). However, the potential and participation of women in the labor force have not been fully utilized. Despite some progress, the economic opportunities and participation of women remain low due to male-oriented development projects and the exclusion of women from the formal employment sector. This often leaves women in unpaid and demanding domestic work, with limited access to training, credit services, and new technologies. As a result, women remain underrepresented in the economic sector of their countries.

Sustainability requires a balance of economic, social, and environmental factors. Neglecting any one of these dimensions can lead to imbalanced and unsustainable outcomes. For example, focusing solely on environmental concerns without considering economic and social factors can make progress unaffordable, while focusing only on economic growth without considering environmental and social impacts can lead to the degradation of natural resources. Furthermore, neglecting the social dimensions of sustainability, such as gender equality, can result in green transitions only benefitting a select few, often men (Stevens 2010).

Green economy initiatives, which aim to create more environmental-friendly economies, must also consider fundamental social needs such as job quality, income equity, and gender equality. Research by Stevens (2010) in Sweden shows that women are more likely to seek information on sustainable consumption choices—purchasing green products, eating organic food, and recycling at a higher rate than men. Similarly, research indicates that Japanese women are willing to pay more for sustainable products, indicating a greater concern for environmental issues than men. Nhamo (2013) argues that the transition to a green economy involves shifting toward more resource-efficient economic and societal structures. It recognizes that economies have multiple goals that need to be achieved through a variety of actions. Therefore, there is no one-size-fits-all approach for nations making the green economy transition. To promote gender mainstreaming within environmental sustainability and climate change within a green economy, it is important to implement gender-specific data, indicators, budgeting, and monitoring. Additionally, the green economy policy framework must create sustainable jobs, not only to promote social and gender equity but to achieve environmental justice and improve the well-being of both women and men. In regions such as Africa and MENA, where there are significant challenges in terms of inequality, poverty, and unemployment, the green economy has the potential to bring significant improvements. Nhamo and Mukonza's (2020) research from South Africa suggests that there are many opportunities for women in the green economy and environmental sectors. They found that 63.5% of respondents agree that the environment enables women's participation in decision-making and in negotiations related to environmental/green economy issues. Those who believed that there were opportunities for women also believed that women were empowered to occupy decision-making roles in all spheres of governance, including the environmental sector. In Africa and the MENA region, the green economy—particularly in the agricultural sector—can be instrumental in empowering women and youth, especially in rural areas.

8.3 Empowering Women: Policy and Incentives

Women's Economic Empowerment (WEE) is an important priority for economic and national security. Donors and implementers understand that women are essential for creating sustainable livelihoods, reducing poverty, and promoting peace. Research shows that women tend to invest a larger proportion of their income in their families (typically up to 90%, compared to 30 to 40% for men). This investment is often directed toward improving the well-being of their children. As a result, the families of working and entrepreneurial women tend to be better nourished, healthier, and more educated. Studies also suggest that when an empowered woman participates in peace negotiations, the resulting agreement is 35% more likely to last at least 15 years. Additionally, women own nearly 10 million of the world's small and medium enterprises (SMEs) and provide jobs for nearly 80% of the SME workforce. Therefore, increasing the profitability of women-owned businesses can create more job opportunities.

Implementers recognize that simply promoting WEE (whether at the population or household level) is not sufficient. To achieve significant change, it is necessary to create an inclusive environment for women in business. This includes policy reforms that expand women's access to economic resources and opportunities, such as those developed by women entrepreneurs, which can provide a basis for collective action and problem-solving. Scholars also acknowledge that empowered women often play a key role in driving reform movements. Women's empowerment seeks to address the status and rights of women. It involves recognizing and utilizing women's full potential and abilities in all areas of life. Empowerment is a multi-dimensional process that enables women to make strategic decisions in their lives. The United Nations Charter, specifically Articles 2 and 23, prohibits discrimination and promotes gender equality, including the right to equal pay for equal work. The UN's Core Guiding Principles also reflect this commitment to equality between men and women in all aspects of human endeavor.

8.4 Empowering Women in the Digital Age

Digitalization has the potential to increase women's participation in labor markets, financial markets, and entrepreneurship. Women's labor is less likely to be replaced by machines than men's, and they have an advantage in the digital age when they possess social skills, higher education, and advanced digital literacy. However, many countries still have barriers and obstacles to prevent women from fully participating in the digital age. Governments need to act to remove these barriers and provide better access to new digital technologies for women in order to achieve gender equality. Policies are necessary to ensure that women can fully participate in—and benefit from—the opportunities presented by the digital age. Without policy interventions, the gender gap in technological and digital access may continue to undermine future gains for women. Recommendations for policy actions have been made to address these challenges, including providing universal and affordable internet access, increasing female digital literacy, encouraging more women to pursue higher education, and promoting web-based entrepreneurship for women. While some governments have taken steps toward these goals, progress has been slow. A study by Sorgner et al. (2017) suggests that the G20 should also act in the following five areas:

1. The G20 should establish an early warning system to monitor potential negative effects on gender equality, particularly with regard to changes in female employment prospects. This system could be triggered by a cross-country initiative. Additionally, the G20 should not only focus on the impact of technology on the female workforce in developed countries, but also conduct research on the potential effects in emerging and developing countries, which may be more vulnerable to digitalization.

2. The G20 should launch an initiative to redesign existing government programs in developing and emerging economies in order to promote women's economic and digital inclusion. This can be done by restructuring social welfare programs to specifically target women as beneficiaries, through financial transfers to their families or communities. By doing this, not only will women's economic inclusion be enhanced, but the effectiveness of the program will also be improved by leveraging women's strengths in terms of their social and family responsibilities. Additionally, incorporating digital technologies more extensively for payment, management, or monitoring can further enhance women's digital inclusion and reduce program delivery costs.
3. The G20 should take more active steps to encourage women to acquire higher education and advanced digital skills. This is necessary as digitalization is likely to create jobs that require social, emotional, and cognitive intelligence, with skills such as creativity and critical thinking developed through higher education. Additionally, proficiency in digital skills will be increasingly important in the digital age. Women often have stronger social skills than men, which they can use to their advantage. However, in emerging economies, there are significant gender gaps in education and digital literacy that may prevent women from accessing high-profile jobs in management, STEM, and entrepreneurship. To address this, many G20 countries have implemented initiatives such as "hackathons," coding workshops, and mentoring programs to promote digital literacy and encourage girls and young women to pursue careers in STEM fields.
4. The G20 should support the creation of high-quality online platforms that provide training and resources for novice female entrepreneurs and connect them with experienced entrepreneurs. Digitalization is creating new opportunities for entrepreneurship, but women often lack the skills and role models needed to recognize and pursue these opportunities. Online platforms can be an effective way to disseminate entrepreneurial skills to women and connect them with mentors and role models, particularly in regions where access to these resources is limited.
5. The G20 should support the development of innovative web-based instruments that improve access to financial capital for female entrepreneurs. High-quality digital platforms that connect female entre-

preneurs with female investors, such as "angel investors," venture capital investors, or equity crowdfunding, should be promoted. Innovative risk assessment methods that rely on big data, rather than traditional forms of securities, should also be utilized to help overcome the lack of collateral available to women entrepreneurs. Such methods, like the Goldman Sachs' 10,000 Women initiative, can utilize online transactions and behavior histories to determine creditworthiness and play a key role in providing financial support to women-led businesses.

8.5 Barriers to WEE and Key Lessons

The Center for International Private Entreprise (CIPE's) approach to reform involves creating new policies and removing existing barriers to women's economic empowerment (see Fig. 8.1). As a private sector policy reform leader, the CIPE focuses on identifying and addressing gaps and silences in existing policies and leveraging existing policy assets to support women entrepreneurs and women-owned businesses. These barriers may include formal policies, such as legal restrictions on women's economic rights and

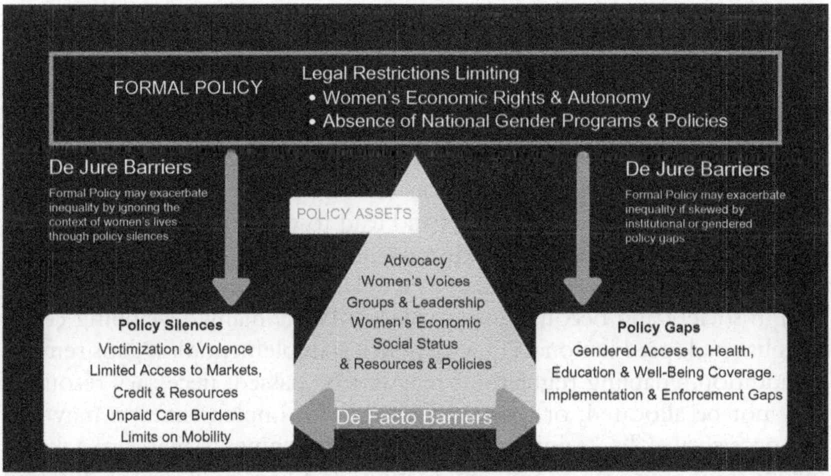

Fig. 8.1 CIPE summary of the policy barriers affecting WEE (Langley and Penumetcha-Huerter 2019)

autonomy, and/or the lack of national programs and policies promoting gender equality. Obstacles may also include informal barriers such as political silence and policy gaps that impede women's economic advancement.

Based on CIPE's experience in designing and implementing WEE programs in over 60 countries, five key lessons have been identified to support effective evidence-based policy reform.

Use a Gender Lens

Gender-neutral policies assume that women are autonomous and free individuals and often fail to account for the fact that men and women lead different lives. As a result, they don't often effectively empower women. In contrast, reform that takes a "gender lens" and is locally driven takes proper consideration of the different ways power is expressed and experienced by men and women, including "power over" (hierarchies, control, discrimination), "power to" (individual agency), "power within" (personal assets and self-confidence), and "power with" (women's groups and collective action). A gender lens approach also considers how women may approach things differently or experience them differently in a given society, as well as factors such as violence against women, legal and policy barriers, or acts of discrimination.

Address Policy Gaps

Policies that intend to treat men and women equally can still be hindered by institutions that do not include women, resulting in a gender gap in policies. Additionally, poor policy design, bureaucracy, and lack of enforcement can impair the implementation of these policies. These policy gaps can further exacerbate inequalities and lead to greater disparities between men and women's overall health, education, and well-being. Structural disadvantages, such as low literacy rates, can limit women's ability to function in society and become entrepreneurs. While many developing countries have adopted national equality policies, implementation gaps remain. In addition, enabling regulations may not be passed, necessary resources may not be allocated, or those responsible for implementation may lack the necessary skills, training, or interest. Governments may also take an inconsistent approach to implementation, and gaps may vary at different levels of government or in different regions, depending on local cultural norms.

Address Policy Silences

Policies that do not account for the unique life contexts of women can perpetuate inequalities through political silence. Factors like gender-based violence, unpaid care work, restricted mobility, and limited access to markets, credit, and resources make it more difficult for women to participate in the political and economic spheres of their communities and countries and can contribute to their marginalization.

Build on Women's Assets

Personal and business assets provide resilience for emerging entrepreneurs and help them overcome challenges. Personal assets may include self-reliance and a willingness to take risks; business assets may include membership in organizations that improve access to institutions, provide a support system for entrepreneurs, and allow for collective action.

Promote Business Resilience

Business resilience is crucial for community resilience. In order for communities to recover and thrive after disasters and other crises, businesses must be able to continue to create jobs and serve their communities. It is important, therefore, to ensure that women entrepreneurs have market-driven business plans and strategies, as well as access to resources, to boost their resilience.

Achieving these five keys are crucial if we are to level the playing field for female entrepreneurs, economically empower women, and enable political reform. Entrepreneurs, both men and women, have a passion for innovation and problem-solving and are adept at bringing their own resources to the political table. Women's equal participation in the economy provides valuable support to policymakers (including elected women) and strengthens free markets and democratic institutions by organizing and amplifying the voices of the private sector.

8.6 Women's Empowerment: Comparisons of Africa and Asia

Women Entrepreneurship and Empowerment in Africa

In Africa, uniquely in the world, the number of women entrepreneurs is higher than that of men. Despite this, data on African women's entrepreneurship is limited (the Global Entrepreneurship Monitor only covers 19 out of 54 African countries for the period 2012–7), and infrastructure to support entrepreneurs—such as accelerators, incubators, venture capital funds, and so on—is also limited. Figure 8.2. presents the estimated value of women's entrepreneurship in Africa in USD billion in 2016.

Women's entrepreneurship significantly impacts the continent, not only in terms of value creation but also in the daily lives of the population. Women in Africa create their own jobs to address market failures that prevent them from accessing employment, unlike in OECD countries where most women hold wage jobs. Data supports this trend, with the difference in female entrepreneurship rates between African countries and the OECD estimated at 17 percentage points. An analysis of startups created by

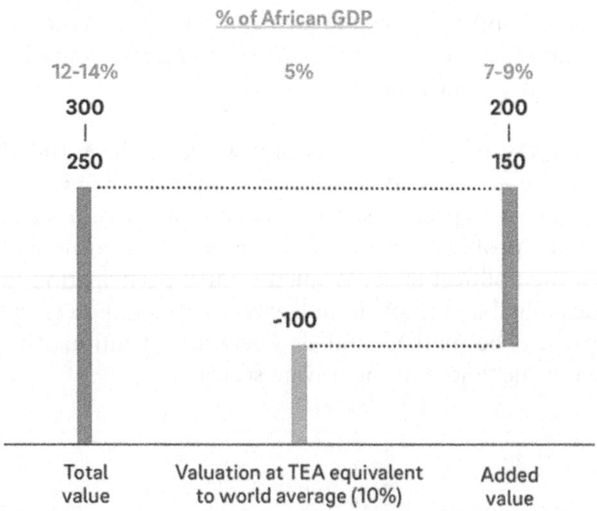

Fig. 8.2 The value of female entrepreneurs in Africa in 2016 (Berger 2018)

women in Africa shows that they mainly operate in the informal sector. Entrepreneurs are typically mothers who take care of large families and start businesses in countries with poor business infrastructure and low education rates. They start businesses out of necessity, rather than desire. These businesses are mainly in sectors such as agriculture, energy, education, and health and have a positive impact on society. To address the challenges faced by women entrepreneurs in Africa, key solutions need to be implemented. These include increasing access to capital, encouraging women to enter male-dominated fields, and providing relevant training. Women entrepreneurs often lack access to assets and capital, which affects their ability to expand their businesses and access medium-sized loans. This can be addressed by giving women more control over assets, eliminating loan collateral, and using alternative methods (such as psychometric testing) to measure loan repayment readiness. Encouraging women to enter traditionally male-dominated fields through mentoring programs and providing relevant information can also help. Furthermore, providing training that focuses on fostering proactive attitudes and entrepreneurial skills can lead to significant increases in business revenues, as seen in Togo where women witnessed a 40% increase in their revenues after completing the training. This type of training has been so successful in Togo that it has been adopted by eight other nations.

Women Entrepreneurship and Empowerment in Asia

In Central Asia, there is a 69.1% gender gap, which researchers suggest will take around 151 years to close. The three highest-ranked countries in this region according to the Global Gender Gap Index of 2021 are Moldova (78.8%), Belarus (75%), and Georgia (73.1%); whilst Azerbaijan (68.7%), Tajikistan (66.3%), and Turkey (63.9%) sit at the bottom. Six out of the ten countries in the region have seen an improvement in their scores, with three countries making particular progress: Armenia (69.8%, +2.5 percentage points), Moldova (78.8%, +2 percentage points), and Kyrgyzstan (70%, +1.9 percentage points) (see Fig. 8.3).

Challenges faced by women in business include lack of financial access, often due to low financial literacy, and a lack of relevant skills training. Improving financial inclusion for women is crucial, as it ensures their access to useful and affordable financial products and services that meet their needs. Despite a global increase in financial inclusion and the number of adults with bank accounts, gender inequality in bank account

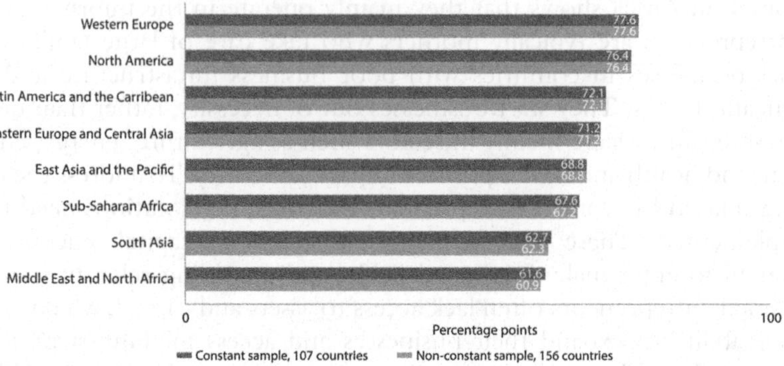

Note: Population-weighted averages for the 156 economies featured in the Global Gender Gap Index 2021 and the 107 economies featured in all editions of the index, from 2006–2021.

Fig. 8.3 Gender gaps by region (%) (ADB Institute, Policy Brief, 2021)

ownership persists. While 72% of men have a bank account, the percentage for women is not mentioned. Financial inclusion is intended to benefit the general public, particularly underserved groups and small and medium-sized businesses (SMEs). However, in practice, the application processes for these services may be too complex for the target population, such as rural women living in poverty, to understand and follow. The Asian Development Bank Institute (ADBI) has invested $22.1 million in projects that promote women entrepreneurs through knowledge-sharing campaigns, training programs, and trade finance. At the national level, governments have partnered with banks to create loans specifically for women to support their entrepreneurial ventures. In the Philippines, for example, it is now easier for women to get business loans through improved online application processes and the waiving of collateral-requirements for female-owned businesses. Additionally, the government provides financial literacy training. In Malaysia, local non-governmental organizations act as intermediaries to bridge the gap between accessible money in rural communities. Local business owners can use the funding to open provision stores and hire local women who are eligible for the funds to shop there for their needs. This eliminates the hassle of opening bank accounts/going to ATMs and improves women's access to inclusive finance. Systems for fostering gender-responsive procurement must be implemented to achieve gender equality in business and encourage women's entrepreneurship. Such systems must include the use of gender-disaggregated data and the alignment of definitions for precision and a shared understanding of its application.

8.7 Concluding Remarks and Recommendations

This chapter analyzes the key factors needed for transitioning to a green economy within the context of women's empowerment. It also provides a foundation for increased focus on training and development for women entrepreneurs in the green economy. It makes a significant contribution to the literature by systematically examining the impact of digitalization on performance and effectiveness within the context of women's empowerment. Additionally, it highlights that sustainable development is a political concept that emphasizes good governance and requires gender parity. Achieving sustainable development involves balancing economic, environmental, and social goals. Policies and plans at all levels must include a commitment to gender equality and the advancement of women's rights and capabilities. This requires addressing gender inequalities and discrimination and including women in productive roles, collective action, and decision-making. Women's empowerment involves creating an enabling environment for women without discrimination and strengthening their ability to control their own lives and make meaningful contributions to society. It requires concerted efforts and commitment from governments, civil societies, and the corporate sectors. Additionally, the equal participation of men and women in decision-making and implementation ensures a fair distribution of resources. Governments should remove barriers to women's participation in the workforce and provide incentives to increase their representation. The role of digitalization and technologies such as the Internet of Things are also discussed in terms of improving social well-being and reducing the impacts of climate change.

References

ABD Institute. 2021. Women's economic empowerment in Asia. ADB Institute. Policy Brief. N°2021–8. https://www.adb.org/sites/default/files/publication/758581/adbi-brief-women-economic-empowerment.pdf.

Alvarez, L. M. 2013. *From unheard screams to powerful voices: a case study of Women's political empowerment in the Philippines.* In: 12th National Convention on Statistics (NCS) EDSA Shangri-la Hotel, Mandaluyong City October 1e2, 2013. https://www.scribd.com/document/394343985/From-Unheard-

Screams-to-Powerful-Voices-a-Case-Study-of-Women-s-Political-Empowerment-in-the-Philippines

Roland Berger. 2018. Women in Africa entrepreneurship: a path to women empowerment. https://www.rolandberger.com/publications.

Braun, P. 2010. Going green: Women entrepreneurs and the environment. *International Journal of Gender and Entrepreneurship* 2 (3): 245–259. https://core.ac.uk/download/pdf/213009278.pdf.

Healy, N., and N. Malhotra. 2009. Myopic voters and natural disaster policy. *American Political Science Review* 103: 387–406. https://doi.org/10.1017/S0003055409990104.

Hechavarria, D.M., Ingram, A., Justo, R., and Terjesen, S. 2012. *Are women more likely to pursue social and environmental entrepreneurship?* Global women's entrepreneurship research. Diverse settings, questions and approaches, pp.135–151. https://www.academia.edu/23285435/Are_women_more_likely_to_pursue_social_and_environmental_entrepreneurship

Langley, B., and Penumetcha-Huerter, S. 2019. Empowering women through policy reform: Five key lessons. CIPE insight from Empowering Women Through Policy Reform: Five Key Lessons - Center for International Private Enterprise (cipe.org).

Mangin, C. 2014. How Women entrepreneurs are contributing to the green economy. https://seed.uno/articles/blog/how-women-entrepreneurs-are-contributing-to-the-green-economy.

Nhamo, G. 2013. Green economy readiness in South Africa: A focus on the national sphere of government. *International Journal of African Renaissance Studies - Multi-, Inter- and Transdisciplinarity.* 8 (1): 115–142. https://www.tandfonline.com/doi/abs/10.1080/18186874.2013.834628.

Nhamo, G., and C. Mukonza. 2020. Opportunities for women in the green economy and environmental sectors. *Sustainable Development* 28 (4): 1–10. https://onlinelibrary.wiley.com/doi/abs/10.1002/sd.2033.

OECD. 2008. Gender and sustainable development: Maximising the economic, social and environmental role of women. In: A report to the United Nations Commission on Sustainable Development (UNCSD) https://www.oecd.org/social/40881538.pdf

Page, Nanette, and Cheryl E. Czuba. 1999. Empowerment: What is it? *Journal of Extension* 37 (5) https://eric.ed.gov/?id=EJ594508.

Smith, S., and K. Musango. 2015. Towards connecting green economy with the informal economy in South Africa: A review and way forward. *Ecological Economics.* 116: 154–159. https://core.ac.uk/download/pdf/37437853.pdf.

Sorgner, A., Bode, E., and Krieger-Boden, C. 2017. The effects of digitalization for gender equality in the G20 economies. Women 20 study. https://www.econstor.eu/handle/10419/170571.

Stevens, C, 2010. Are Women the key to sustainable development? Sustainable Development Knowledge Partnership (SDKP), USA. https://www.bu.edu/pardee/files/2010/04/UNsdkp003fsingle.pdf.

Sumathi, K., T.S. Anuradha, and S.B. Akash. 2014. Green business as a sustainable career for Women entrepreneurs in India: An opinion survey. *Advances in Management* 7 (5): 46–47. https://search.proquest.com/openview/d89ebe1e96addaf17d3c29ce55fb8348/1.pdf?pq-origsite=gscholar&cbl=2030322.

United Nations Development Programme. 2012. *Gender and climate change: Capacity development series*. New York: UNDP. https://www.uncclearn.org/wp-content/uploads/library/undp118.pdf.

USAID. 2016. *Engendering utilities: Improving gender diversity in power sector utilities*. USAID's Office of Energy and infrastructure, energy division within the Bureau of Economic Growth, education and environment in Washington. Retrieved from https://www.usaid.gov/sites/default/files/documents/1865/Engendering-Utilities.pdf Last Accessed 30 Oct 2016).

Webb, J. 2015. Gender dynamics in a changing climate: how gender and adaptive capacity affect resilience. *CARE* Gender and Adaptation Learning Brief. https://www.researchgate.net/publication/338337753_Opportunities_for_women_in_the_green_economy_and_environmental_sectors.

CHAPTER 9

Considering Gender Lens in the STEM Field

Amira Kaddour, Imen Gmach, and Sinda Elghoul

9.1 Introduction

In the knowledge economy, jobs require new skills. In fact, increasing the interoperability of concepts and theories and how individuals use digital skills, knowledge, and technology calls for a new educational model, which is driven by education 4.0. paradigm (Hong and Ma 2020).

A. Kaddour (✉)
National School of Advances Sciences and Technologies, University of Carthage, Amilcar, Tunisia

Includovate, Tunis, Tunisia
e-mail: amira.kaddour@enstab.ucar.tn

I. Gmach
Imam Abdulrahman Bin Faisal University, Dammam, Saudi Arabia
e-mail: ihgmach@iau.edu.sa

S. Elghoul
CRISTAL Laboratory, GRIFT Research Group National School of Computer Science (ENSI) Manouba, Manouba, Tunisia
e-mail: sinda.elghoul@ensi-uma.tn

K. Drucza et al. (eds.), *Centering Gender in the Era of Digital and Green Transition*,
https://doi.org/10.1007/978-3-031-38211-6_9

Education needs to prepare people to support societal progress, teach the STEM basics, while not neglecting the ethical aspects of the new digital world in terms of civic and environmental responsibilities, especially in the face of artificial intelligence. According to Norton et al. (2020), digital transformations require a change in the way organizations work since it is driven by cutting-edge digital technologies and cutting-edge commercial strategies. That requires more than simply putting technological solutions in place, an agreement between organizational, human, and digital aspects. For Mahlow and Hediger (2019) digital transformations create new models and skills through the intentional and deep use of digital technologies that must serve human development. To promote diversity and economic inclusion, education must promote these values (Herring 2009; Muchiri and Ayoko 2013; Reagans and Zuckerman 2001). STEM fields are at the core of these transformations and need to drive inclusive transformations by considering the needs of women. In the knowledge economy, jobs require new skills. In fact, according to Hong and Ma (2020), we may need a new educational model to ensure we capacitate ourselves to take on the challenges of the digital transition. Education needs to teach the STEM basics, while also considering the ethical aspects of STEM technologies. It must prepare people to support societal progress, while engendering civic and environmental responsibilities (especially in the face of threats from artificial intelligence). According to Norton et al. (2020), digital transformations not only require changes in skills and changes in education, but changes in the way that organizations work. To reap the greatest benefits, organizations cannot simply put cutting-edge technological solutions in place, but must weigh and balance human, digital, and organizational desires and demands. Mahlow and Hediger (2019) suggest that we must intentionally use digital transformations to serve human development and promote diversity and economic inclusion, and that STEM education must promote these values (Herring 2009; Muchiri and Ayoko 2013; Reagans and Zuckerman 2001). STEM fields are at the core of digital transformations and need to drive inclusive transformations by considering the needs of women.

9.2 The Role of Education in Digital and Green Transitions

Education plays a crucial role in digital and green transitions by capacitating individuals with the necessary skills and knowledge to participate in—and contribute to—these transformations. That includes digital training in skills such as coding, data analysis, and digital literacy, as well as education on sustainable practices and the principles of the green economy. Education can also play a role in raising awareness and promoting engagement in digital and green initiatives, helping to drive transformations at both the individual and societal levels. Specific skills are required for accessing jobs in the digital economy, including coding, data analysis, and digital literacy. Coding is in fact the foundation of the digital economy, since it helps create the software, websites, and apps that power the internet and digital devices. Organizations are also increasingly looking for employees who can analyze and make sense of the vast amounts of newly generated data so as to inform decision-making in science, business, marketing, and elsewhere. Furthermore, digital literacy allows individuals to navigate and use digital technologies effectively and is essential for jobs in e-commerce, digital marketing, and customer services, while digital skills allow individuals and organizations to collaborate effectively using tech tools and platforms and are essential for remote work, virtual teams, and job opportunities in project management. In addition, educating individuals on sustainable practices, renewable energy, and environmental science can help facilitate access to jobs, as well as have a significant impact on society and the environment. Promoting awareness and understanding of digital and green technologies and practices, and explaining how they can be used to create a more sustainable future, will encourage critical thinking and problem-solving skills, which are essential for tackling the complex issues related to the digital and green transitions. STEM education includes subjects such as physics, chemistry, biology, computer science, engineering, and mathematics. It provides students with the knowledge and skills to develop new technologies and solutions that can help create a more sustainable future. For example, engineers and scientists can develop new renewable energy technologies, while computer scientists and data analysts can develop new tools and methods for monitoring and analyzing environmental data. Moreover, STEM education can foster an innovative mindset in students, encouraging them to think outside the box, come up with new ideas and solutions to sustainability challenges. It can equip students with the analytical, critical thinking, and problem-solving skills necessary to tackle complex issues, such as climate change and environmental degradation.

STEM education can help equip the workforce with the necessary skills and knowledge to participate in the green economy, as well as make informed decisions about the sustainability issues that affect their lives. This empowers individuals, particularly women and underprivileged communities, and in turn promotes sustainable economic growth and development.

9.3 STEM: A Continually Evolving Field

The STEM field is continually developing and innovating. That requires our ongoing commitment to acquiring and upgrading the necessary skills and capacities. Teachers and educational planners have a key role to play in this, not least in designing STEM curricula to prepare students for jobs in the field. Teachers can integrate technology into all their lessons (regardless of their primary subject), in order to capacitate students with the skills and mindsets that can effectively prepare them for the Fourth/Fifth Industrial Revolutions. Scientific progress depends on strong support from society (financial and non-financial), as is shown by key studies (Ben-David 1971), and there are strong connections between science and societal conditions (Xie et al. 2015). Students need to adopt a critical and scientific mindset if they are to contribute to innovation and decision-making in the competitive twenty-first century (Aberilla et al. 2021). The evolution of learning theory is constantly progressing and developing (Hanisch and Eirdosh 2020), and we will need to discover and evolve our teaching strategies to help students understand developments in their daily lives. Students' performance is often attributed to their personal experiences and environment, and strengthening learning modalities will allow them to build new knowledge and ideas (Orongan et al. 2019; Aberilla et al. 2021). While STEM education is not a new concept, schools are changing their approach to it. Every day, students are surrounded by technology. Instead of focusing on teaching students how to use technology, technology should be at the center of student learning activities to maximize learning outcomes. Moving to a skills-based, technology-supported curriculum allows students to develop specific skills and a mindset that will prepare them for future jobs as well as to meet the challenges of a rapidly changing world. Dasgupta (2019) demonstrate that traditional education skill sets are insufficient for the demands of our new world, and that timely interventions are needed to improve skills. Online education should be supplemented by online assessment. Educators must also upskill themselves by making good use of the current online teaching tools. Many studies have found that when educators use an integrated approach, female participation increases and higher education outcomes improve. Educators must therefore not only change their subject-based mindset but also modernize and intelligently tutor themselves. The STEM tree formulation (see Fig. 9.1) is intended to provide a model solution for integrating various subjects (Deák et al. 2020). The technological age is characterized by rapid developments in the STEM fields. This has increased demand for individuals with STEM skills, so that businesses, organizations, and

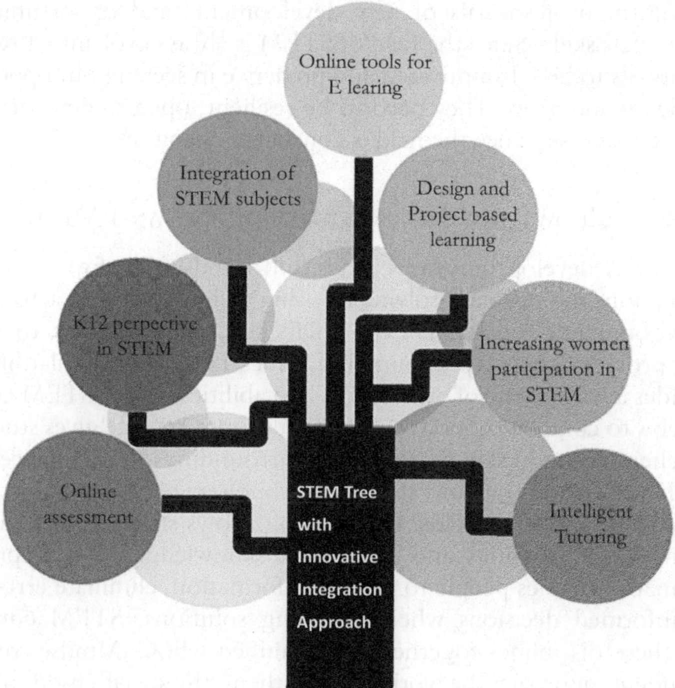

Fig. 9.1 STEM Tree Modelling, an innovative integration approach (Deák et al. 2020)

governments can keep up with changes and make use of the new technologies. There has been significant growth in STEM careers in recent decades and this trend is expected to further increase in the future. The rapid pace of technological advancements means that new tools, techniques, and technologies are constantly emerging, and professionals need to continually upgrade their skills in order to stay competitive and relevant in their respective industries. As the field of STEM is interdisciplinary, employees who are experts in their own discipline, while also having a broad understanding of other fields, are likely to have an advantage. Continuously upgrading skills through formal education—that is classes, workshops, conferences, seminars, professional training programs, development courses, advanced degrees, and so on—can also help inform professionals of STEM advances. Networking with other professionals also helps inform professionals of new developments and opportunities to upgrade their skills. Since the field of STEM is always evolving, it requires professionals to be self-motivated and proactive in seeking out opportunities to learn and grow. They need to be resilient, open to new ideas, and willing to take risks since the field is constantly changing.

9.4 Growth of Innovative Abilities and Mindsets

Students can develop innovative skills and mindsets to foster creativity, collaboration, and problem-solving by using technology in an active learning environment. Skill sets are the ability to complete a task to solve a specific problem, such as programming, data science, and simulation skills. It includes a wide range of experiences and abilities. Every STEM component helps to complete a well-rounded education. Science gives students a comprehensive understanding of their surroundings and enhances their research and critical thinking abilities. Technology prepares young people for careers in high-tech fields. Engineering allows students to hone their problem-solving abilities and apply their knowledge to new projects. Mathematics enables people to analyze information, eliminate errors, and make informed decisions when designing solutions. STEM education brings these disciplines together into a unified whole. Mindset refers to how students approach the world around them, the social-emotional skills they employ, and the methods they use to solve problems, such as computational thinking. To examine the importance of STEM education and its impact on tomorrow's needs, it is worth positioning it within Kondratieff's global "waves of innovation" framework (see Fig. 9.2). In this Kondratieff

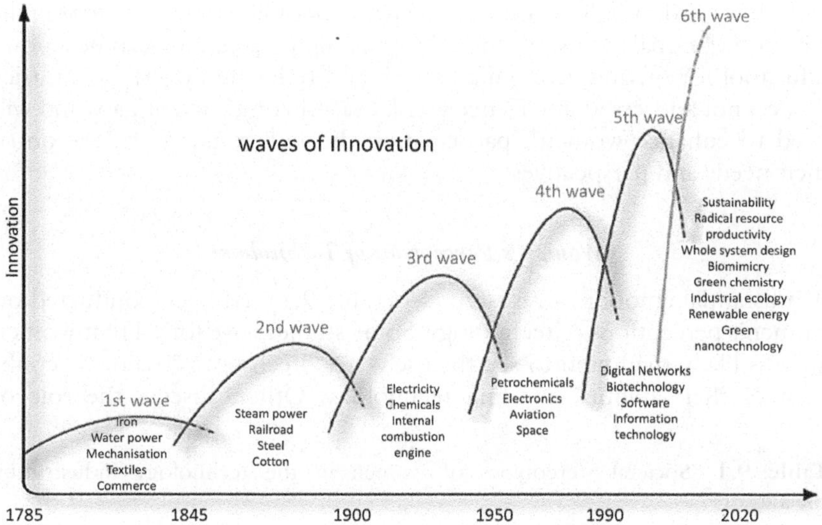

Fig. 9.2 Kondratieff cycles Vs innovation (Nair and Paulose 2014)

presents the 40- to 60-year cycles, or megatrends, that have impacted our socioeconomic life from the beginning of the nineteenth century. Today, we are in the sixth wave, where digital transformations including those in the STEM sector are leading new economic models.

9.5 STEM Field: Barriers and Obstacles for Women

According to a 2017 report by UN Women:

> *"despite large efforts made over the past decades to narrow the gender gap in STEM education, major inequalities still persist. Socio-economic, cultural, and other obstacles still prevent female learners from completing or benefiting fully from good quality education of their choice in many situations."* Moreover, *"women's exclusion from the digital world has shaved $1 trillion from the gross domestic product of low- and middle-income countries in the last decade—a loss that will grow to $1.5 trillion by 2025 without action. Reversing this trend will require tackling the problem of online violence, which a study of 51 countries revealed 38% of women had personally experienced"* (UN Women 2022).

In low- and middle-income countries, economic models are more reliant on traditional forms of activity. Yet the digital economy can be a powerful tool for driving economic growth and reducing poverty. Countries that do not address women's needs risk exacerbating current gaps and will need to enhance women's participation through a full consideration of their needs and perspectives.

Women's Perception of Technology

A substantial amount of research (see Table 9.1) has been conducted on women's perceptions of technology. Some studies have found that women are less likely than men to see themselves as "tech-savvy" or to be confident of their abilities in using technology. Others discuss the role of

Table 9.1 Societal stereotypes of women in the technology industry by the authors

Sample studies	*Key findings of studies*
Dholakia et al. (1994), NTIA (1999), Badagliacco (1990), Bimber (2000), Faulkner (2001)	Women came late to the technology sector (less interested)
Survey conducted by TechGirlz, in partnership with Drexel University's LeBow College of Business (2019)	86% of girls said their parents had encouraged their interest in technology, with fathers more supportive than mothers (94% vs. 85%). Friendship (47%) was the main motivator for girls of tech-related activities; career potential was a far less important incentive (mentioned by only 10%). However, parents considered that career potential (33%) was the biggest motivator for girls to learn about technology. All sixth-grade girls expressed an interest in technology (100%), compared to 88% of eighth-grade girls.
21st session of the UN Commission on Science and Technology for Development (2018)	"Negative stereotypes about women and girls studying STEM subjects are among the impediments to an inclusive world where hi-tech solutions solve global problems. Digital skills are not optional in today's tech-savvy world."
Katila (2022)	In gaming, new "tech-savvy" identities challenge current stereotypes' by focusing on coding as a "normal" (rather than exceptional) activity for women; these also shift "attention away from the machine to the person and what they can accomplish with the machine."

families and their impact on girls' perceptions. Such perceptions can be due to a variety of factors, including a lack of exposure to technology in early life, a lack of female representation in the tech industry, and societal stereotypes about women and technology.

Some studies have shown that women are less likely to pursue careers in technology-related fields and that they are underrepresented in technology companies. This can lead to a lack of role models and mentors for women in technology, which can further reinforce the perception that technology is a "male" field.

Other studies have found that women tend to have a more critical and nuanced view of technology and its impact on society. They are more likely to consider its potential negative consequences, such as privacy concerns and impacts on social interactions. Finally, it is critical to note that the above perceptions can vary depending on culture, education, and socioeconomic background and that not all women have the same perceptions of technology.

Stereotypes

There are several social stereotypes that impact women's access to STEM—the most common are included in Table 9.2.

9.6 Concluding Remarks and Recommendations

Women make up half the world's population but are vastly underrepresented in the STEM sector. This needs to be addressed if we want to build inclusive societies and achieve the equality goals detailed in the 2030 Agenda for Sustainable Development. The lack of female representation in STEM is a complex issue with many contributing factors, including social stereotypes, lack of role models for girls and discrimination against women, lack of support for women in the workplace. We need to create more opportunities for women to enter the sector and provide support for those who are already in the field, including through mentorship and scholarship programs, and flexible work arrangements. We also need to create more awareness and encourage more girls to pursue STEM fields. According to the World Bank (2021), improving women's place and empowerment in STEM is a critical strategy for achieving the SDGs.

Table 9.2 Stereotypes that impact women's access to STEM

Stereotypes	*Assumptions*	*Studies*
Women are not as committed to their careers as men	**Assumption:** Women are more likely to put family and other responsibilities ahead of their work. **Corrective:** Studies have shown that men and women have similar levels of commitment to their careers. Women are more likely to be the primary caretakers of children and may face more barriers to pursuing careers in STEM than men.	Relational cohesion theory See: Sloan (2017)
Women are not as interested in STEM as men	**Assumption:** Women lack the same level of passion and drive. **Corrective:** Research has shown that girls and women have similar levels of interest in STEM as boys and men.	See Chauke (2022).
Women are not as competent as men	**Assumption:** Women are more likely to make mistakes or fail. **Corrective:** Research has shown that men and women have similar levels of competence in STEM fields.	See: Hammond and Rubiano-Matulevich (2020)
Women are not as good at math and science as men	**Assumption:** Because women are less good at STEM, there are fewer women in the fields. **Corrective:** Studies have shown that men and women have similar abilities in math and science.	See article: "The Myth of the Male Math Brain," AAUW.

Women play a determining role in the evolution of digital transitions, and understanding their needs, improving their presence, and progressing their careers are critical both for the wellbeing of women and for global inclusion and equality goals.

References

Aberilla, O., Salic-Hairulla, M., Orbita, R.R., and Torres, M.A.J. 2021. University Students' Acceptance of Evolution: Basis for STEM-based Instructional Design. *International Journal of STEM Education for Sustainability*, 1 (1):33–44e-ISSN. 2798–5091. https://doi.org/10.53889/ijses.v1i1.3.

Badagliacco, J.M. 1990. Gender and race differences in computing attitudes and experience. *Social Science Computer Review* 8 (1): 42–63. https://psycnet.apa.org/record/1990-26888-001.

Ben-David, J. 1971. The Scientist's Role in Society: A Comparative Study. Prentice-Hall, 1971–207.

Bimber, B. 2000. Measuring the gender gap on the internet. *Social Science Quarterly* 81 (3): 868–876. https://citeseerx.ist.psu.edu/document?repid=rep1&type=pdf&doi=2a9cc6656dac376cf4a413bf29f4549cdd80b868.

Chauke, T.A. 2022. Gender differences in determinants of students' interest in STEM education. *Social Sciences* 11: 534. https://doi.org/10.3390/socsci11110534.

Dasgupta, P. 2019. Time and the Generations: Population Ethics for a Diminishing Planet New York, NY: Columbia University Press.

Deák, V., Nemecskó, I., Végső, T., and Bódi-Schubert, A. 2020. A koronavírus járvány hatása a magyarországi pénzforgalomra 2020-ban (Impact of the coronavirus pandemic on Hungary's payment turnover in 2020). Special article (szakmai cikk), Magyar Nemzeti Bank. https://www.mnb.hu/letoltes/a-koronavirus-jarvany-hatasa-a-magyarorszagipenzforgalomra-2020-ban.pdf. Downloaded: 17 March 2022.

Dholakia, R.R., N. Dholakia, and B. Pedersen. 1994. Putting a byte in the gender gap. *American Demographics* 16 (12): 20.

Faulkner, W. 2001. The technology question in feminism: A view from feminist technology studies. *Women's Studies International Forum* 24 (1): 79–95. https://www.sciencedirect.com/science/article/abs/pii/S0277539500001667.

Hammond, A., and Rubiano-Matulevich, E. 2020. Myths and misperceptions: Reframing the narrative around women and girls in STEM'. https://blogs.worldbank.org/education/myths-and-misperceptions-reframing-narrative-around-women-and-girls-stem.

Hanisch, S., and Eirdosh, D. 2020. Educational potential of teaching evolution as an interdisciplinary science. *Evo Edu Outreach* 13 (25). https://doi.org/10.1186/s12052-020-00138-4.

Herring, C. 2009. Does Diversity Pay?: Race, Gender, and the Business Case for Diversity. *American Sociological Review* 74 (2): 208–224. https://doi.org/10.1177/000312240907400203.

Hong, C., and Ma, W.W.K. 2020. Introduction: Education 4.0: Applied Degree Education and the Future of Work. In: Hong, C., & Ma, W. (Eds.), *Applied Degree Education and the Future of Work. Lecture Notes in Educational Technology*. Springer, Singapore. https://doi.org/10.1007/978-981-15-3142-2_1.

Katila, S. 2022. How 'tech-savvy' women are changing masculine dominance in the video game industry. https://wearetechwomen.com/how-tech-savvy-women-are-changing-masculine-dominance-in-the-video-game-industry/.

Mahlow, C., and Hediger, A. 2019. Digital transformation in Higher Education—buzzword or opportunity? *eLearn Magazine* 2019 (5): 13. https://doi.org/10.1145/3329488/3331171.

Muchiri, M., and Ayoko, O.B. 2013. Linking demographic diversity to organisational outcomes: The moderating role of transformational leadership. *Leadership & Organization Development Journal* 34 (5). https://doi.org/10.1108/LODJ-11-0086.

Nair, S., and H. Paulose. 2014. Emergence of green business models: The case of algae biofuel for aviation. *Energy Policy* 65: 175–184. https://www.researchgate.net/publication/259513304_Emergence_of_green_business_models_The_case_of_algae_biofuel_for_aviation.

Norton, A., Shroff, S., and Edwards, N. 2020. Digital transformations, an entreprise architechture perspective. PublishNation Limited. https://www.researchgate.net/publication/341297551_Digital_Transformation.

NTIA (National Telecommunications and Information Administration). 1999. Falling through the net: Defining the digital divide. A report on the Telecommunications and Information technology gap in America. Revised. U.S. Government Printing Office, Superintendent of Documents. http://www.ntia.doc.gov/ntiahome/fttn99/contents.html.

Orongan, M.J.Q., Nabua, E.B., Barquilla, M.B., Buan A.T., Inutan, E.N., and Yuenyong, C. 2019. Cognitive attributes, physical and psychosocial aspects of learning environment: Its relationship to learners' chemistry achievement. Journal of Physics: Conference Series, Volume 1340, International Annual Meeting on STEM Education (I AM STEM). https://doi.org/10.1088/1742-6596/1340/1/012068.

Reagans, R.E., and Zuckerman, E.W. 2001. Networks, Diversity, and Productivity: The Social Capital of Corporate R&D Teams. *Organization Science* 12 (4): 502–517. https://doi.org/10.1287/orsc.12.4.502.10637.

Sloan, M.M. 2017. Gender differences in commitment to state employment: The role of coworker relationships. *Public Personnel Management* 46 (2): 170–187. https://doi.org/10.1177/0091026017702612.

World Bank. 2021. Infographic: The Equality Equation: Advancing the Participation of Women and Girls in Science, Technology, Engineering and Mathematics (STEM). https://www.worldbank.org/en/news/infographic/2021/04/19/the-equality-equation-advancing-the-participation-of-women-and-girls-in-science-technology-engineering-and-mathematics-s.

UN Women. 2022. Gender snapshot report 2022. https://www.unwomen.org/sites/default/files/2022-09/Progress-on-the-sustainable-development-goals-the-gender-snapshot-2022-en_0.pdf.

Xie, Y. Fang, M. and Shauman, K. 2015. STEM Education. *Annual Review of Sociology* 41: 331–357. https://doi.org/10.1146/annurev-soc-071312-145659.

CHAPTER 10

Gender-Based Violence: Basics and Evolution

Amira Maazouz and Amira Kaddour

10.1 Introduction

In his 1755 *Discourse on the Origin and Basis of Inequality Among Men*, Jean Jacques Rousseau distinguished between two kinds of inequalities: one that was "natural" or "physical" (i.e. established by nature and concerning different ages, strengths of body, and qualities of mind or soul) and one that was "moral" or "political," that is, the result of privileges (such as being honored, being powerful, or being obeyed) enjoyed by some to the detriment of others (Rousseau 1996). However, these two main categories of inequalities can harbor a multitude of inequalities such as gender inequality.

A. Maazouz
Institute of Accounting and Administration, University of Gafsa, Gafsa, Tunisia

A. Kaddour (✉)
National School of Advances Sciences and Technologies, University of Carthage, Amilcar, Tunisia

Includovate, Tunis, Tunisia
e-mail: amira.kaddour@enstab.ucar.tn

K. Drucza et al. (eds.), *Centering Gender in the Era of Digital and Green Transition*,
https://doi.org/10.1007/978-3-031-38211-6_10

10.2 Gender-Based Violence: The Basics

The concept of "structural, gender-based power imbalances" refers to the ways that societal systems, institutions, and norms disproportionately benefit men and disadvantage women. These imbalances can manifest in various ways, including through economic, political, and social inequalities. These power imbalances create conditions that make women and girls more vulnerable to abuse, exploitation, and violence in part because they tend to have fewer economic resources and less social support than men. They are also more likely to experience sexual violence because of societal norms around gender and sexuality that often blame victims, excuse perpetrators, and normalize/excuse the mistreatment of women and girls. Additionally, women and girls may be specifically targeted in situations of armed conflict because of their gender.

Violence: A "Chameleon" Concept

Some commentators consider violence a "chameleon" because it changes its appearance depending on the environment in which it is found. Women are disproportionately victims of violence. (The word "victim" derives from the Latin term *victima* which incorporates the notion of "sacrifice". A "victim" was, in essence, "a living being offered as a sacrifice to the gods".) In a legal sense, though, a victim is defined as a person "who has personally suffered harm, as opposed to the person who causes it, known as the author of the harm" (Guinchard and Montagnier 2007). People may be direct or indirect victims, with the latter referring to "third parties suffering material or moral damage as a result of the damage caused to the victim" (Guinchard and Montagnier 2007). In short, the concept of being a "victim" is inseparable from the concept of being harmed (Idriss and Abbas 2011). Violence can take various forms. It may be physical, moral, or economic (Mazeaud et al. 1991), with moral violence consisting "of moral pressure" (KANT 1994; Carbonneir 2000), and the concept of economic violence emerging in the 1970s (David and Pugsley 1985) and being recognized by English courts for the first time then: a key jurisprudential paradigm shift in our understanding of violence (Majdii 2009). The concept of violence by its elastic character has created a fog on the level of its delimitation.

In his book *The History of Violence*, Jean Claude Chesnais considers that "Violence is not one but multiple, constantly changing concepts, according to places, times, circumstances, and even environments. In this modern era, it has been enriched with new forms, constantly more complex and more abundant" (Chesnais 1981).

Violence and Gender: What Interaction Between Both Concepts?

The concept of gender had a thorny way before getting recognized by the international community,[1] although it constitutes "a useful concept for rethinking the non-discrimination rights" as demonstrated today by feminist studies (Fondimare 2014). The interaction between the concepts of "violence" and "gender" has given birth to a full-fledged concept namely gender-based violence which is broadly acknowledged as "a serious human rights abuse." Indeed, the topic of "GBV: basics and evolution" is marked by a certain conceptual subtlety that requires exploring the different interactions between the concept of "gender" and the new faces of violence. By revealing the development gaps induced by the social relations of sex between males and females, the notion of gender-based violence has provided a pedestal for developing a broader sense of inequalities. Gender studies exceeded the binarity of the two sexes and focused on the concept of gender-based violence. According to article 3 of the Council of Europe's Convention on Preventing and Combating Violence against Women and Domestic Violence, GBV is defined as "violence that is directed against a woman because she is a woman or that affects women disproportionately." However, concepts of gender-based violence have varied across time, countries, and cultures. Indeed, the concept of discrimination based on the social roles recognized for women and men has fed a frightening rise in inequalities, with economic and social inequalities visible across many fields and sectors including education, health, and employment (Ben 2016).

In the digital era, the scope of gender-based violence has expanded to include digital practices, with Buturuga Roumanie, n° 56,867/15. Cyberviolence was first explicitly defined and recognized by the European

[1] Article 3 of the Council of Europe Convention on Preventing and Combating violence against Women and Domestic Violence (signed in Istanbul in 2011) defines "gender" as "the socially constructed roles, behaviors, activities, and attributes that a given society considers appropriate for women and men."

The term "gender" is borrowed from the Latin word *genus* which means "socially-constructed sex" and refers to "the feminine and masculine roles, behaviors, activities, and attributes that a given time considers appropriate and that is determined by a set of social and cultural norms."

Court of Human Rights in 2020, with the Court viewing it as an aspect of violence against women and girls that can take various forms, including invasions of privacy (i.e. taking, sharing, and manipulating data/images, including intimate ones).

10.3 New Forms of GBV in the Digital Era

According to the International Humanitarian Organization (2020) Plan which sampled 14,000 women, 58% of those surveyed had experienced online harassment, with half saying they had been more harassed online than on the street. Digitalization can ease women's life and facilitate inclusion, but it can also bring new forms of violence which can even have a greater impact than more traditional forms of violence. This following section will present and analyze these issues.

From Off-Line to OnLine Violence

In the twenty-first century, social networks such as Facebook, Twitter, Instagram, WhatsApp, Snapchat, and TikTok are widely used forms of communication. However, they can also provide platforms for cyberviolence especially against women (Hare and Oliversi 2021). To date, some international and European legal frameworks have been put in place to prevent and combat violence against women. Some of these contain binding provisions, such as the Istanbul Convention, which addresses all forms of violence against women, and the Council of Europe on Cybercrime (Budapest Convention), which addresses cybercrime. Other legal tools and instruments include the Recommendation on Preventing and Combating Sexism, adopted in 2019 by the Council of Europe Committee of Ministers, which contains the first ever internationally agreed definition of sexism and specifically addresses sexist hate speech online. In addition, the European Commission against Racism and Intolerance (ECRI) aimed to combat hate speech and hate crime in its 2019 General Policy Recommendation.

The Different Faces of Digital Violence

Cyberspace is a breeding ground for gender-based cyberviolence. Trolls and other aggressors can create hostile online environments that aim to shame and intimidate women. Although bullying can take place both face-to-face and online, online bullying leaves a digital footprint, which can be hard to erase but also useful as proof of the abuse. Online violence against women is an extension of off-line violence against them. It can take many forms, including direct or indirect threats of violence (physical or sexual), bullying, revenge porn, or doxing (i.e. disseminating documents online that contain personal information without that person's consent.).

An International Legal Framework to Address GBV

Cyberviolence accelerated from 2020, largely due to enforced lockdowns during the COVID-19 pandemic. The international community has responded by adopting the Istanbul Convention, which provides an important legal framework to prevent and combat digital violence against women. The Convention presents an arsenal of legally binding standards and covers gender-based violence in article 3. Since then, the Council of Europe's Group of Independent Experts on Action Against Violence Against Women and Domestic Violence (GREVIO 2021) has adopted a general recommendation to combat harmful acts against women and girls that are committed online or facilitated by technology. This is complemented by other tools including the Council of Europe's Budapest Convention on Cybercrime, which offers a comprehensive set of legally binding standards aimed at criminalizing cyberviolence, ensuring the collection of evidence, and establishing cross-border and international cooperation in combating online violence against women. In addition, the Council of Europe's Recommendation on Preventing and Combating Sexism and ECRI's General Policy Recommendation also aim to combat hate speech, as detailed also above.

Criminalizing: Cyberviolence is a dangerous phenomenon that needs to be anchored in criminal laws. Under French law, cyberviolence is considered "online harassment" (see article 222-33-2-2 of the French Penal Code) and the French Data Protection Agency (CNIL–Commission Nationale de l'Informatique et des Libertés - France) is tasked with addressing digital attacks, whether on social networks, or in photo-sharing applications.

10.4 GBV and Ecosystem Health: Interaction

There are complex interactions between gender and environmental issues (climate change, natural disasters, resource depletion, etc.), which can exacerbate gender-based violence. For example, women and girls may be disproportionately affected by climate change because they often have less access to resources and decision-making power, but may be more likely to be caring for family members. This can make them more vulnerable to direct physical violence as well as indirect violence through displacement, loss of livelihoods, and increased poverty. Resource depletion and scarcity can also increase the risk of gender-based violence. Women and girls may be disproportionately affected by competition for resources such as land, water, and food. Moreover, women and girls are also more likely to be the victims of violence (e.g. when they are forced to travel long distances to gather water and firewood), which puts them at particular risk of harassment and assault in insecure regions. Addressing the complex interplays between gender and environmental issues is thus critical to safeguarding women and ensuring safe and healthy environments for them.

10.5 Advanced Technologies to Tackle GBV

Legal standards and ethical recommendations can only go so far: technologies must also adopt preventive measures to protect women's access to digital platforms and technological tools. To ensure that technology is safe and accessible for women, it is important for developers and designers to incorporate preventive measures and ethical considerations at the earliest stages of design. Key steps to ensuring women's rights are protected include conducting user testing and research with diverse groups of women, including those from marginalized communities, to ensure that the technology meets their needs (in terms of accessibility, usability, and safety) and addresses potential challenges, biases, and barriers that may impact women's use.

Technology can exacerbate violence against women, but can also help prevent and address gender-based violence not least by providing support for survivors. Hotlines and online resources, for example, can provide information, services, and support for survivors, and early warning systems can also help monitor and report incidents of gender-based violence.

Advanced technologies have the potential to play a significant role in addressing gender-based violence. For instance:

- The use of surveillance and monitoring technologies—such as facial recognition, audio/video monitoring, and GPS tracking—can help detect and prevent gender-based violence by monitoring individuals or high-risk locations. For example, facial recognition technology can help identify individuals who have a history of committing gender-based violence and can alert authorities when they are detected in certain locations. Audio and video monitoring can detect and record incidents of violence in real time, which can help provide evidence for prosecution. GPS tracking can be used to monitor the movements of known offenders as well as track the location of victims in order to assist them. By helping detect and prevent gender-based violence, these technologies can create safer environments for women and other at-risk individuals. However, it's important to note that these technologies must be implemented with a strong focus on privacy and data security and must be used within the boundaries of the law.
- Mobile applications can be designed to inform survivors how to access medical care, legal assistance, and other resources. Emergency response systems—such as panic buttons, smart home devices, or wearable devices—can provide survivors or at-risk individuals with quick and discreet ways to contact emergency services or their support network. These technologies can also help collect data to improve response and support for GBV survivors, as well as identify patterns and trends in gender-based violence that can be used to inform the development of targeted prevention strategies.
- Big data analytics help process and sort large amounts of information from diverse sources (police reports, hospital records, social media, etc.) to help identify patterns and trends in incidents of gender-based violence. This information can be used to identify high-risk areas, high-risk populations, and common characteristics of offenders, all of which can inform the development of effective and well-targeted prevention strategies. For example, data analytics can help identify key times of day, locations, or circumstances that are associated with increased risk of gender-based violence, and preventative actions can thus be effectively implemented (i.e. increasing police presence in high-risk areas or providing education/awareness programs to high-risk populations). Data analytics can additionally evaluate the effectiveness of existing prevention strategies and help identify areas where additional resources are needed, ensuring that resources are

directed where they are most needed and preventative actions are the most effective.

- Digital counseling and therapy technologies (such as virtual reality, video conferencing, and chatbots) can provide mental health support to survivors of gender-based violence. Virtual reality can be used to provide exposure therapy to individuals who have experienced trauma, which in turn can help reduce symptoms of anxiety and depression. Video conferencing can provide therapy sessions to individuals who may have difficulty accessing in-person support, such as those who live in remote areas or with mobility issues. Chatbots can be used to provide emotional support and information to survivors 24/7. These technologies can provide a convenient and accessible way for survivors of gender-based violence to access the mental health support and other key services. It can help to reduce the stigma associated with seeking therapy and increase access to mental health services for individuals who may otherwise not have access to them. It's important to note that these technologies should be used in conjunction with traditional therapy, and should be approached with caution, as they may not be appropriate for everyone. Chatbots should also be designed to recognize signs of distress and provide appropriate referrals to professionals. Privacy and data security must also be properly considered.
- Education and awareness technologies, such as social media and mobile apps, can be used to raise awareness about gender-based violence, provide education, and connect survivors with resources and support. Social media, for example, can be used to disseminate information about gender-based violence and to raise awareness about key issues. It can also be used to provide education on healthy relationships and consent and to connect victims with resources and support. Mobile apps can also be used to provide educational resources, such as information on how to access medical care, legal assistance, and other resources. They can also help victims contact emergency services or contact their support network. By providing education and raising awareness, these technologies can help reduce the incidence of gender-based violence by changing the attitudes and behaviors that contribute to GBV. Additionally, by connecting victims with resources and support, these technologies can help to empower survivors and improve overall GBV responses.

10.6 Concluding Remarks and Recommendations

New forms of technology have led to new forms of gender-based violence, including cyberstalking and revenge pornography. Cyberstalking uses technologies like social media and email to stalk, harass, or threaten individuals. Revenge pornography uses the non-consensual sharing of sexually explicit images or videos, often with the intent of humiliating or harming the individual depicted, or blackmailing them, by threatening to share images or videos unless the individual performs specific actions. These forms of online violence can be particularly harmful and difficult to escape, since perpetrators may have constant access to victims through technology. They can also be difficult to prosecute because they often cross international boundaries and may be difficult to detect. Online forms of violence are not separate from off-line violence, but rather an extension of it. Since both forms of violence are interconnected, it's important to address them both in a holistic manner and to work together to prevent and respond to gender-based violence in all its forms. Women are disproportionately affected by online violence and are more likely to be targeted because of their gender and because digital platforms are not often designed with consideration for the safety and security of women. Online violence can have a severe impact on women's mental and physical health, their relationships, and their ability to participate in online activities. It can also lead to economic consequences, such as loss of income or employment. Women have less power to defend themselves against online violence and often cannot access the same resources and support as men (such as legal representation). They may additionally face stigma and discrimination when seeking help These issues can make it especially difficult for women to report incidents of online gender-based violence and to access support and resources to address them.

Creating a safe and inclusive online environment for women requires the concerted efforts of different stakeholders—including governments, tech companies, civil society organizations, and the media. Online gender-based violence will not be adequately addressed without policies and practices that promote gender equality and respect for all human rights. The voices and participation of affected communities, especially women and girls, should be central to this process.

References

Ben Achour S. 2016. Violences à l'égard des femmes :lois du genre. Publications d'EuroMedDroits,Tunis, p. 23 https://euromedrights.org/fr/publication/tunisie-les-violences-a-legard-des-femmes-lois-du-genre/

Carbonneir J. 2000. Droit civil: les obligations. Tome 4, *22ème édition refondue* Paris PUF p 103.

Chesnais JC. 1981. "Histoire de la violence", R. Laffont, coll. PLURIEL", Paris, p. 11.

David R., and Pugsley D. 1985. Les contrats en droit anglais, 2ème édition. Paris LGDJ, p. 196

Fondimare Elsa. 2014. "Le genre, un concept utile pour repenser le droit de la non-discrimination", dossier thématique : Révolutions et droits de l'Homme (I)Aspects épistémologiques et conceptuels. Penser les révolutions conceptuelles des droits de l'homme. Revue des droits de l'homme, N°6. https://doi.org/10.4000/revdh.755

Grevio. 2021. General Recommendation No. 1 on the digital dimension of violence against women. Adopted on 20th of October 2021 www.coe.int/convention violence.

Guinchard G., and Montagnier G. 2007. Lexique des termes juridiques, 16ème éd., Dalloz, p.673.

Hare I., and Oliversi A. 2021. Analyser les cyberviolences au prisme du genre. Presses universitaires de Lorraine, 2021/2 n°40, p. 321, 324. https://journals.openedition.org/questionsdecommunication/27108

Idriss M.M.. and Abbas T. 2011. Honour, violence, women and Islam. Routledge, p. 102, 103, 115

Kant E. 1994. *Métaphysique des mœurs*. Tome 2 Doctrine du droit, doctrine de la vertu; Paris GF- Flammarion, p 129.

Majdii W. 2009. La violence en matière contractuelle, Mémoire pour l'obtention du master en droit des affaires. FSJPST, p.14 en MORETEAU (O), "*Droit anglais des affaires*" Paris DALLOZ DELTA 2000, p. 306.

Mazeaud, H., Mazeaud, L., Mazeaud, J., and Chabas, F. 1991. Leçons de Droit Civil Tome II. Premier Volume: Obligations: théorie Générale, 8ème édition par François CHABAS, MONTCHRESTIEN. p. 182.

Rousseau J.J. 1996. Discours sur l'origine et les fondements de l'inégalité parmi les Hommes. Librairie Générale, Commentaires et Notes par Gérard MAIRET, p. 75

PART IV

Monitoring Economic Development Toward Just Transition

CHAPTER 11

The Benefits of Considering Gender in Economic Development

Sujata Ganguly and Elena Nikolova

11.1 Introduction

This chapter argues that gender equality and economic development are tightly intertwined, and that gender aspects should be an essential consideration for economic policies, projects, and programs. Gender relations affect all aspects of human life, and achieving gender equality is important for reasons of equity, equality, and efficiency. This chapter provides a critical review of the existing literature on gender in economic development, a relationship which is important for both academics and policymakers to understand. A growing body of literature shows that gender inequality is very costly for societies. According to a 2015 study, gender-based discrimination in formal/informal laws, social norms, and practices restrict

S. Ganguly (✉)
Includovate, New Delhi, India
e-mail: sujata.ganguly@includovate.com

E. Nikolova
Zayed University, Dubai, UAE
e-mail: Elena.Nikolova@zu.ac.ae

K. Drucza et al. (eds.), *Centering Gender in the Era of Digital and Green Transition*,
https://doi.org/10.1007/978-3-031-38211-6_11

women's rights and access to opportunities and costs the global economy an estimated US$12 trillion (McKinsey Global Institute 2015). Additional research shows that violence against women costs the global economy US$1.5 trillion, equivalent to 2 percent of the global gross GDP (Puri 2016).

While some have argued that gender equality will naturally follow economic development and growth, the existing evidence contradicts this line of thinking. Although many countries have achieved remarkable poverty reduction and growth trajectories, women's positions often have not improved. For instance, over the past 25 years India has experienced rapid economic growth, urbanization, declining fertility rates, and rising educational levels among women. However, nationally women's labor force participation declined by 25 percent between 1983/4 and 2011/2 (Lahoti and Swaminathan 2016).

This chapter starts by outlining a conceptual framework, arguing that gender equality *drives* economic development and growth via six main channels: (1) productivity and innovation; (2) corporate governance; (3) better serving female customers; (4) increased spending on household nutrients, health, and housing; (5) children's wellbeing; and (6) education. The framework also discusses how these channels interact with women's inclusion in the green economy, as well as in sustainable and inclusive growth more broadly. The conceptual framework is presented in Sect. 11.2 of this chapter.

Section 11.3 applies this conceptual framework to review the literature linking gender and economic development in four broad domains: (1) employment and social protection; (2) unpaid work and the care economy; (3) the green economy and gender inequality; and (4) COVID-19 as a window of opportunity for women in the green economy. The discussion emphasizes how women's "double burden" of unpaid and underpaid work limits their economic opportunities and incomes (both in the global South and in the global North), concluding that gendered ideas about women's roles and capacities may be to blame. The last part of this chapter focuses on the links between gender inequality and the green/brown economies and points out that the post-COVID-19 recovery should be used as a window of opportunity to initiate a development model that encompasses gender inclusiveness in these domains. Finally, we offer our concluding remarks.

11.2 Methodology

The chapter is informed by a desk review and takes into consideration the inclusion and exclusion criteria as depicted in Table 11.1 below.

Table 11.1 Inclusion criteria for the analysis

Criteria	*Inclusion criteria*
Data sources	Reports, journal articles, grey literature
Research perspective	Gender equality and economic development are tightly intertwined; gender aspects should be an essential consideration for economic development.
Search terms	Green economy, gender equality, economic development, employment, social protection, unpaid work, care economy, COVID-19

11.3 Conceptual Framework

In this section, this chapter develops a conceptual framework explaining how gender equality and economic development influence each other. In particular, the framework emphasizes the interaction between SDG 5 (gender equality), SDG 3 (good health and wellbeing), and SDG 4 (quality education). In the subsequent sections, this chapter applies this framework to discuss how gender affects economic development in four broad domains with reference to the green economy.

Does Economic Growth Lead to Gender Equality?

Do gender equality and economic development go hand in hand? One group of scholars and policymakers argue that as national incomes increase, gender equality will follow naturally (the idea being that "a rising tide lifts all boats"). Advances in technology and the associated evolution of labor markets, it is said, will increase the bargaining power of women, while attitudinal changes will ensure that discrimination against women—in the workplace and elsewhere—is eradicated. As a result, policies should solely target economic growth, rather than gender equality. However, several pieces of evidence suggest that the impact of economic growth on gender empowerment is not clear-cut. First, even in developed countries today, women who are equally qualified continue to earn less than men at all levels of qualification (Olivetti and Petrongolo 2016). Motherhood is penalized equally harshly in both developed and developing countries.

Recent research (Nizalova et al. 2016) estimates that the motherhood "wage penalty" (the difference in wages between mothers, and women who have no children) is 33 percent in Germany, 19 percent in Ukraine, 10 percent in Russia and 15 percent in Hungary (though 0 percent in Finland and Sweden). Second, sex ratios remain skewed in favor of boys in many countries, largely due to the availability of sex-selective abortion (see Guo et al. 2016 for evidence on China and India; Brainerd 2013 for evidence on Armenia and Azerbaijan). Third, there is still a big gender gap in political participation. In 2017, only around a quarter of members (25.7 percent) in national parliaments around the world were women (Inter-Parliamentary Union 2021). Finally, it may matter where growth comes from: growth driven by natural resources, such as oil or other minerals (as is the case in much of Central Asia), may reduce the number of women in the labor force, thus encouraging the development of institutions and laws which perpetuate the low status of women (Ross 2008).

The Effect of Gender Equality on Economic Development and Growth

Is it then possible that gender equality may actually drive development and growth? There are at least three reasons to believe this may be the case. First, gender inequalities in education have been found to reduce annual per capita growth rates by about 0.4 to 0.9 percent. This is largely due to decreases in overall human capital (i.e. through the selection of less-able men over more-able women) and the declining returns of investing in male education (Klasen 2002). Excluding women from education and the labor force may stifle innovation. Second, gender diversity at the organizational level can have a positive effect on productivity and overall economic growth. For instance, different groups have access to different information (some of which may not overlap), and exposure to different views can lead to creativity and innovation (Nikolova 2017). As Lagarde and Ostry (2018) point out, there is considerable microeconomic evidence that women and men bring different skills and perspectives to the workplace, such as attitudes toward risk and collaboration. Ostry et al. (2018) built a model to estimate the gains to GDP from reducing gender inequality around the world and demonstrated that closing the gender gap could increase GDP by an average of 35 percent in the bottom half of the least gender-equal countries. Four fifths of these gains come from adding workers to the labor force, while one fifth is due to the gender diversity effect on productivity.

Gender-related barriers to employment are equivalent to an estimated 4 percent tax on female labor in the average country in Europe and Central Asia and rise to an estimated 53 percent tax on female labor in countries across the Middle East and North Africa. In South Asia, East Asia and the Pacific, gender inequality is effectively equivalent to tax rates on female labor of 32 percent and 11 percent respectively (Ostry et al. 2018). These figures are staggering and highlight the extreme inefficiency of shutting women out of the labor market. There is also evidence that companies perform better when they have more women on their boards or in senior management positions. Companies which employ female managers may be better suited to serving consumer markets dominated by women, and more gender-diverse boards could enhance corporate governance (Elborgh-Woytek et al. 2013). In the public sector, women as policymakers prefer policies that better reflect their own priorities, including child health and family nutrition, as well as policies that increase their bargaining position within the household, improve their rights in case of divorce, and raise their chances of access to the labor market (Duflo 2012). Women's presence in parliament also decreases corruption (Jha and Sarangi 2018).

Women's empowerment may also positively affect a variety of family outcomes. Women's education has been found to be positively correlated with child welfare (particularly health), and research in the United States demonstrated that giving women the right to vote decreased infant mortality (Miller 2008). Income or assets in the hands of women are associated with greater improvements in child health and larger household expenditure on nutrients, health, and housing (Duflo 2012). Moreover, children who are better nourished and more educated are likely to be more productive adults. This link between women's empowerment and children's wellbeing is highly relevant for SDG 5 ("Achieving gender equality and empowering all women and girls"). Women's empowerment is an important aim in itself, but these studies also show that it may also have positive, long-term spillover effects on children and families.

Figure 11.1 visually summarizes the links between gender equality and economic growth and development as discussed in this section.

The Green Economy and Gender Equality

Gender equality, women's economic empowerment, and inclusive green growth are all essential for achieving sustainable economic development and poverty alleviation. Our current economic development model is ecologically unsustainable and inequitable (DCED 2019), with climate

Fig. 11.1 How gender equality increases economic development

change impacts disproportionately effecting women. That said, actions to address climate change are often agnostic to gender differences, and thus perpetuate the exclusion of women and social injustice (Schulz 2021). The Global Green Growth Institute (GGGI) sees the transition to green growth as an opportunity to accelerate gender equality and women's empowerment by leveraging women's roles in green growth and climate action as decision-makers, entrepreneurs, workers, and consumers (GGGI 2021). The relationship between gender equality, women's economic empowerment, and green growth is further set out in the 2030 UN Agenda for Sustainable Development, and also clearly established in other international policy frameworks, including the United Nations Framework Convention on Climate Change (UNFCCC) Gender Action Plan (see Box 11.1).

Box 11.1 Green Growth and Gender in International Policy Frameworks

The 2030 Agenda for Sustainable Development

"We affirm that green economy policies in the context of sustainable development and poverty eradication should [...] enhance the welfare of women [...] continue efforts to strive for inclusive, equitable development approaches to overcome poverty and inequality [...] mobilize the full potential and ensure the equal contribution of both women and men."

UNFCCC—Decision—/CP.23—Establishment of a Gender Action Plan

"Gender-responsive climate policy continues to require further strengthening in all activities concerning adaptation, mitigation and related means of implementation [...] Parties should, when taking action to address climate change, respect, promote and consider their respective obligations on human rights as well as gender equality [...] The Conference of the parties adopts the gender action plan [...] with a view to advancing towards the goal of mainstreaming a gender perspective into all elements of climate action."

The Rio Declaration—Principle 20

"Women have a vital role in environmental management and development. Their full participation is therefore essential to achieve sustainable development."

The Paris Agreement

"Parties should, when taking action to address climate change, respect, promote and consider [...] gender equality, the empowerment of women and intergenerational equity."

The Sendai Framework for Disaster Risk Reduction 2015–2030

"More than 1.5 billion people have been affected by disasters [...] with women, children and people in vulnerable situations disproportionately affected. Empowering women [...] to publicly lead and promote gender-equitable and universally-accessible responses, recovery, rehabilitation and reconstruction approaches is key."

The Beijing Declaration and Platform Agenda

"Eradication of poverty based on sustained economic growth, social development, environmental protection and social justice requires the involvement of women in economic and social development, equal opportunities and the full and equal participation of women and men as agents and beneficiaries of people-centered sustainable development ..."

Source: DCED (2019)

11.4 Discussion

This section applies the conceptual framework developed in Sect. 11.2 to review the linkages between gender and economic development in the domains of (1) employment and social protection; (2) unpaid work and the care economy; and (3) the green economy.

Employment and Social Protection

Women comprise just 40 percent of the global labor force (Kochhar et al. 2017). According to Brussevich et al. (2018), women's participation in the labor market leads to enhanced productivity, economic growth, reduced income inequality, and increased economic resilience. However, women's involvement in economic activities does not by itself necessarily drive inclusive growth, since persistent gendered ideologies continue to perpetuate gender inequality. Women may be subject to sexual harassment and discrimination at work (Elson and Pearson 1981; Corossacz 2018), and working women also often take on the lion's share of unpaid care work within their households, leading to a "double burden" of unpaid and underpaid work that may potentially undermine their wellbeing and productivity (Ilkkaracan 2018). Moreover, due to care responsibilities at home, women may refrain from working full-time or even working at all, thus receiving less wages and fewer benefits than men over their lifetimes (Angelov et al. 2016). Women tend to be paid less than men for doing the same work (Elson and Pearson 1981; Wilson 2015), and also take on a disproportionate amount of precarious and unpaid work, such as domestic work (Salardi 2016; Corossacz 2018). Women dominate the informal sector, especially in the Global South, with little or no access to social security and other benefits, including medical support, maternity leave, unemployment, or disability benefits (Majumdar and Borbora 2013). In short, while it is important to increase women's participation in the labor force, that alone will not allow inclusive growth. So long as gendered notions of women's capabilities exist, and the devaluation of women's work remain unaddressed, inclusive growth will not be achieved.

Discrimination and Exploitation Although the incorporation of women in the labor market is often deemed economically beneficial by neoliberal feminists, gender discrimination and the exploitation of women's labor persists (Elson and Pearson 1981; Wilson 2015). For example, under the creation of the New International Division of Labour (NIDL), less developed economies in the Global South were increasingly integrated into the world economy, leading to the creation of employment opportunities, especially in the garment sector. Large fashion industries based in the Global North searched for cheap labor (largely provided by women) and outsourced manufacturing to developing countries, such as Bangladesh (Dey and Basak 2017). The export-oriented, ready-made garment industry is critical for the economy of Bangladesh and other developing "sweatshop economies." Women working in this sector often come from impoverished, rural backgrounds and aren't protected by trade unions that can advocate for better wages and working conditions (Siddiqi 2009; Dey and Basak 2017). In addition, women's work is often categorized as "unskilled" or "semi-skilled," while similar work undertaken by men (e.g. in agriculture) is considered "skilled," meaning that women often receive lower wages than men (Elson and Pearson 1981; Wilson 2015).

The Overrepresentation of Women in the Informal Economy The informal sector has been on the rise in part due to the "gig economy," in which workers are able to work flexibly through digital platforms. While more flexible work schedules can help women better balance paid and care work (Milkman et al. 2021), "non-standard" employment often does not provide fair compensation, union protection, social security benefits, or job stability (Young 2010). While the Indian Ministry of Labour and Employment legally recognized gig workers in 2019, there are still significant barriers to women accessing digital education or skills training, and thus the gig economy in India remains dominated by men (Kasliwal 2020). Care and domestic work are also part of the informal sector and are disproportionately carried out by women, especially women of color (Salardi 2016; Corossacz 2018). For example, while Brazil has seen a substantial increase in its middle class (Neri 2010), gender inequality remains entrenched. More than 90 percent of Brazilian domestic workers are women, half of whom are of African descent (Corossacz 2018). According to Corossacz (2018), the power relationship between white middle-class men and female Afro-Brazilian domestic workers remains profoundly unbalanced due to the intersection of sexism and racism by the former toward the latter. The double discrimination based on gender and race often leads Afro-Brazilian women to have little or no choice but to enter the informal sector through precarious, domestic work. In a study by Gonçalves (2010), a black, female interviewee stated that "unfortunately, paid domestic work is a door into the work market for black women in our country." Moreover, female domestic workers can be subject to gender-based violence (GBV) perpetrated by their employers, sons, husbands, relatives, or friends (Gonçalves 2010; Corossacz 2018). The invisibility and undervaluation of women's paid domestic labor also connects to global inequality and migration, since many low-skilled immigrant women end up in domestic positions (Williams 2010). A qualitative study conducted in Spain—one of the major countries importing migrant workers—documented poor working conditions, low wages, and health hazards for female immigrant workers from countries such as Colombia, Morocco, Romania, and Ecuador (Ahonen et al. 2009). Female migrant workers remain vulnerable to labor exploitation largely due to their immigration status (especially those who are undocumented), the lack of legal regulation in the sector, language barriers, and the non-transferability of their education and vocational training (Ahonen et al. 2009; Murphy 2013).

Social Protection as a Pillar of Inclusive Growth and Gender Equality Social spending should be considered an investment rather than charity, since it contributes directly to gender equality and empowerment. Lack of social protection stifles women's labor force participation. Gender gaps in the labor force perpetuate gender inequalities in social protection, while the informal sector is characterized by a lack of social protection (UNICEF 2021). In order to achieve gender equality in social protection, it is crucial to treat women as individual right-bearers, independent of their relationship to a male breadwinner, and to prioritize access to universal social services based on the needs of caregivers and receivers. Additionally, policy solutions must be implemented, through universal child allowances, reforms of pension and health systems, and the use of community work programs. These are tool for gender-equitable social protection. Finally, it is essential to consider women's needs and constraints in the design and implementation of targeted social protection programs to ensure substantive equality for all (UNICEF 2021).

Unpaid Work and the Care Economy

The care economy encompasses the goods and services necessary for the physical, social, and emotional care of dependent individuals (including children, the elderly, the ill, and people with disabilities). Studies show that women take on the lion's share of unpaid care work throughout the world, which limits their economic activities and economic development more broadly (Addati et al. 2018; Ilkkaracan 2018). Gendered division of labor ideologies, which permeate many societies, tend to be overlooked and inadequately addressed in economic development programming. This subsection discusses the invisibility and precariousness of women's work (care and unpaid) and examines how women's responsibilities in the household affect their participation in the labor market. The section also discusses policy solutions to address women's unpaid work and examines how such policies can complement global efforts toward inclusive growth.

The Persistent Glass Ceiling for Women Attributed to Their "Double Burden" Even in countries with highly inclusive economies, persistent gender pay gaps remain. A study from Sweden conducted by Angelov et al. (2016) shows that—despite high female labor force participation, parental leave for both men and women, and social protection policies—there remains a glass ceiling for female workers. The authors attribute this to women's greater involvement in child rearing compared to male partners. Moreover, women tend to face barriers when it comes to getting promotion or pay rises due to gendered preconceptions about their capabilities and gender-biased views about their performance (Ridgeway and Correll 2004). The care economy ensures that individuals function at a socially acceptable level of capability, comfort, and safety. However, there are significant gender inequalities when it comes to time spent on unpaid work around the world. Even when men participate in unpaid/care work, they tend to spend significantly less time on domestic chores such as food preparation and cleaning (Alonso et al. 2019). According to UN Women (2018), almost half of productive work time globally is unpaid, the allocation of time between paid and unpaid work is subject to gross gender imbalances, and women work more than men when paid and unpaid work are combined. Additionally, women spend substantially more time on unpaid work than men, regardless of the country.

The Gendered Division of Labor Ideologically driven gendered perceptions often intersect with socio-economic statuses, household structures, and societal institutions. Women in many countries are prohibited from working and confined to private spheres (Kochhar et al. 2017). Women's access to property—such as land and other productive assets—is critical for entrepreneurship, business opportunities, and overall economic growth. Yet women's property rights are restricted by law in some countries, including those in sub-Saharan Africa (Joireman 2008). Such a distinctive gendered division of labor is largely based on the primitive notion of men as productive, and women as reproductive (Rose 1986). Patriarchal gender roles, in short, undermine women's socio-economic status and entrench gender inequalities. Gender discrimination, exploitation, and unequal opportunities can have negative impacts on national economies by undermining productivity and growth (Kochhar et al. 2017). Since gender inequality reduces women's material and overall wellbeing, policy and program interventions should consider the intertwined issues of women's labor and sexual exploitation, the undervaluation of women's work, women's access to fairly paid employment and social protection, property rights, and the burden on women of combining paid, home, and reproductive work.

The Impact of Unpaid Care Work on Women's Economic Empowerment A well-functioning care economy is crucial to both economic and social sustainability. Unpaid work impacts inclusive growth, human development, and gender equality, as well as poverty alleviation, human development, job generation, productivity, the care crisis, and sustainable growth. According to Alonso et al. (2019), the value of unpaid work globally varies from 10 percent to 60 percent of GDP. The unequal burden of unpaid care work on women has a significant impact on their economic empowerment. Women often experience a double burden, with unpaid care work also resulting in "time poverty" which limits leisure time, prospects for better jobs, and longer hours of paid work (Chopra 2015). This phenomenon forces women into home-based or low-skilled jobs, often working from home while simultaneously carrying out household chores which further reduces productivity (Chopra 2015). Such gendered roles are a root cause of inequalities in the labor market, resulting in occupational and vertical gender segregation, a gender employment gap, the clustering of women in a few "feminine" job sectors, and a gender pay gap (UN Women 2018). The gendered division of labor that results from unpaid care work perpetuates gender stereotypes and biases, with women being perceived as less capable and competent than men due to their domestic responsibilities. Such perceptions often prevent women getting promotions or pay rises, further entrenching gender inequalities in the labor market (UN Women 2018). Furthermore, women's participation in the labor market is largely shaped by their domestic care responsibilities, and in the absence of care leave and affordable quality care services, many women end up having to quit their jobs after childbirth or as family members grow old or fall sick (UN Women 2018). Women's participation in the labor force is negatively impacted by unpaid care work, resulting in lower wages, limited access to education and training, and fewer opportunities for career advancement (Folbre 2018). Therefore, recognizing the importance of unpaid care work is crucial to promoting economic development and achieving gender equality and women's economic empowerment. This recognition can lead to more gender-equitable policies and practices in the workplace and society as a whole (Folbre 2018).

Heteronormativity in the Analysis of Gender and Economic Development While the gendered division of labor is indeed a significant cause of gender inequality, an exclusive focus on women would risk normalizing women as caregivers (Nelson 2016) and assuming that households always contain women and men (Jolly 2011). Non-traditional households, such as single-parent households, are largely overlooked in the current mainstream analysis of the care economy and development programming (Jolly 2011). For instance, in development programming practices, female-headed households are often considered poorer than male-headed households, since men are assumed to be the breadwinners (Jolly 2011). As Jolly argues, female-headed households may consist of single women who choose not to engage in relationships with men. In addition, men who undertake household work either voluntarily or involuntarily (such as men in single-parent households) are often rendered invisible (Nelson 2016). Excluding non-traditional households from analysis and policy making precludes us from understanding how gender inequality and the gendered division of labor can be addressed fully.

The '3R' Strategy as a Framework for Policy Interventions Related to Unpaid Work and the Care Economy The '3R' strategy involves Recognizing, Reducing, and Redistributing unpaid work. This strategy calls for actions that go beyond merely increasing the visibility of unpaid work, to include policies that alleviate women's care burden and more equitably divide caring between men, women, families, and public/market services (UN Women 2018). National perceptions of women's role in society influence the success of the 3R Strategy in each country. For example, Alonso et al. (2019) find that values and perceptions matter when it comes to redistributing unpaid work more equally between men and women. Societies that don't value gender equality in jobs, politics, and education as much (conditional on the level of economic development) have less-unevenly distributed unpaid work. Male unpaid work is also considered higher in societies where more people consider equal rights key to democracy or disagree with statements such as "men make better business executives than women do," "men make better political leaders than women do," or "a woman has to have children to be fulfilled." Several policy interventions have been identified to help better recognize, reduce, and redistribute unpaid work. These include investing in social care infrastructures, time-saving (rural) physical infrastructure, and work-life balance policies, as well as regulating the labor market to promote gender equality. Supporting macroeconomic fiscal policies is also crucial (UN Women 2018).

Green Economy and Gender Inequality: Bridging the Gaps

As discussed in the previous sub-sections, women continue to be "invisible" in certain countries and contexts, and rarely get to exercise their human rights, especially economically. For this reason, attention is given to sustainable industrialization and gender equity/equality in the 2030 Agenda for Sustainable Development, which argues that gender equality and environmental sustainability are inter-connected. In other words, both women and the environment are undervalued in our global systems, while both are essential if we are to see transitions to inclusive sustainable economies (McLean 2019). There is extensive evidence that women's empowerment leads to economic growth and inclusive sustainable development (UN Women 2014). For example, a survey of 21,980 public companies from around the world showed that companies that had 30 percent women in leadership roles had a 6 percent higher net profit margin (Adamy 2016). Implementation of the SDGs requires that stakeholders connect and coordinate their actions rather than work in "silos." While the connections between the green economy and women's economic empowerment is obvious, it needs further elaboration. Although data on gender in the green economy exists at the global level, gender-disaggregated data at the national and provincial levels is critically missing, particularly in low- and middle-income countries. The literature suggests that the obstacles limiting women's participation in the green and brown economies are similar. Some of the key limiting factors influencing women's participation in green economic activities are the prevailing gender pay gap, the unpaid care burden, lack of education and skills (particularly in science, technology, engineering, and mathematics), lack of financial literacy and access to information, discrimination and harassment, and male dominance in key industries (i.e. in energy and emerging green technologies). Women are more likely to rely on agriculture, forestry, and fisheries for income, yet they have limited ownership, access to—and control over—land and other resources. Rural and indigenous women are especially marginalized. The scarcity of natural resources and climate disasters, which are increasingly common in the global South, further exacerbate the care burden and increase violence against women (McLean 2020). However, women tend to have smaller ecological footprints than men and engage in more sustainable behaviors (McLean 2019). Women and men respond to and approach environmental issues differently and have different levels of access to—and control of—natural resources. Yet women are powerful agents of change and can play a crucial role in the greening of economies (McLean 2019). The green economy offers—through the creation of new jobs—many opportunities to change women's participation in the labor market. It provides a conceptual shift as well as a starting point from which to value women's contributions. Engendering the green economy is crucial and has immense benefits for economic efficiency, equity, and environmental sustainability. This allows for targeted actions, innovative solutions, increased flexibility, improved environmental management, effective investment, equitable distribution of resources, empowerment, and poverty alleviation. Women are key actors for environmental protection and possess knowledge, capacities, and capabilities to drive solutions related to sustainability and resource scarcity. Women are, therefore, critical agents of change, and their participation in the green industry and contributions to the green economy are vital for achieving equitable environmental sustainability. Social, environmental, and economic considerations can jointly lead to inclusive green growth pathways and ambitious climate action by integrating gender equality and pro-poor approaches (Schulz 2021). In many developing countries, though unrecognized and undervalued, women—as forest stewards, farmers, natural resource managers, and entrepreneurs—already engage in green economic activities. For example, in India women's management of small green businesses (often based on forest and agricultural products) allow them to participate in the workforce both as laborers and entrepreneurs (Jadhav 2009). While there are concerns and apprehensions that emerging job opportunities in key green industries tend to be male-dominated, there are opportunities to strategize pathways for both women and men to learn new skills and access green job opportunities.

Box 11.2 below summarizes new research on the economic empowerment of women in the green industry in Cambodia, Peru, Senegal, and South Africa.

Box 11.2 Empirical Evidence from a Study on the Economic Empowerment of Women in Green Industries in Cambodia, Peru, Senegal, South Africa (UNIDO 2021)

There is currently limited evidence of women's empowerment and involvement in the green industry. This is problematic given the existing gender inequalities in entrepreneurship roles in the green industry, which are expressed (both overtly and covertly) in discriminatory practices. However, more women are becoming—or aspiring to become—entrepreneurs in the green industry than in conventional industries, and there is a strong perception (among the women interviewed) that there are more equal opportunities for women to progress in the green industry.

The female entrepreneurs interviewed in the four countries reported similar barriers to starting a green business, namely:

- The high costs of business startup
- Lack of awareness about starting a business
- Lack of access to technology
- Lack of understanding of definitions (i.e., what makes a business "green" and what does not)
- The comparative expense of producing green products over those manufactured by "conventional industries"
- The lack of incentives for businesses operating as green industries.

While there are policies on women's economic empowerment and climate change adaptation, there is a general lack of synergy between the two, with a dearth of gender-responsive green industrial policies. The study found that although there are organizations and institutions working to support women entrepreneurs across the four countries, none specifically focus on supporting women in the green industry. This could be attributed to the fact that the green industry is new and not yet well defined across countries.

COVID-19 Impacts and its Window of Opportunity for Women in the Green Economy

As discussed earlier in this chapter, women in the Global South are more likely than men to be involved in the informal sector, where activities have been disproportionately affected by lockdown measures. The COVID-19 pandemic has been a health, social, and economic crisis and had major impacts on family relations, due to the closing of schools and nurseries. Many women have been struggling to balance their professional and personal lives, and women working in the informal sector (such as domestic workers) are particularly vulnerable (UNEP and UN Women 2020). For example, while the number of COVID-19 cases in Cambodia has been relatively small compared to its neighbors (i.e. Indonesia, Malaysia, Philippines, and Thailand), the country's tourism, garment, footwear, and construction sectors have been heavily affected due to declining global demand. (The garment, textile, and footwear sectors employ more than 750,000 workers, most of whom are internal women migrants from remote provinces, see UNIDO 2021.) Women are also among the worst affected by the escalating climate crisis. Even so, to date, they have barely participated in shaping and evolving responses to either crisis (the climate or COVID-19). As gender gradually becomes a more significant factor in climate change mitigation policies, it is critical to learn from global best practices and strategies and develop a robust framework for women in the green economy.

Before the COVID-19 pandemic, the Asia/Pacific region was leading the world in the creation of green jobs, with the support of various investments, stimulus packages, and policy reforms aligned with the SDGs. Efforts were made to encourage and support women's participation in—and contribution to—parts of the green industry, which historically has had limited female representation. However, due to pre-existing gender imbalances, investments in the green economy may not necessarily result in women's equal access to green jobs, nor provide the resources that would position them effectively for opportunities in the green industry. The COVID-19 pandemic has exposed many fragilities in overall economic development, exacerbated existing gender inequalities, and made the plight of the poor and marginalized more visible. Yet it can also catalyze a rethinking and redesigning of approaches to support systemic change. There are growing calls to use the crisis to design appropriate strategies and build more inclusive, sustainable, and low-carbon circular economies. Despite its many challenges, the pandemic also offers the global community a unique opportunity. Women have long been seen as critical agents of post-crisis recovery and investing in gender equality has the potential to stimulate economies and reverse losses in global wealth. The World Economic Forum has highlighted the unique window opportunity for a "Great Reset" of our economies and societies (UNEP 2020).

Box 11.3 Summarizes some of the proposed mitigation measures that can be adapted to spearhead the green economy, achieve gender equality, and address the COVID-19 crisis.

Pandemic-related economic measures can be an avenue to invest in the structural transformations and technological innovations needed for sustainable development. Apart from providing economic opportunities in the short term, such improvements are essential to enhance the overall resilience of societies. Greening recovery measures can provide countries with opportunities to "build back better," combining economic growth *and* achieving environmental goals (OECD 2020).

Box 11.3 Mitigation Measures that Can be Adapted for the Green Industry (CARE 2020)

- Collect and make available data disaggregated by sex and age, as well as data/information on disabilities and at-risk groups.
- Ensure information, messaging, and community outreach on COVID-19 is inclusive and accessible to all and does not reinforce harmful gender stereotypes.
- Develop mitigation strategies that specifically target the economic impact of the pandemic on women and build women's financial resilience.
- Ensure women from diverse backgrounds and from different socio-economic strata can meaningfully engage in structures and processes established for COVID-19
- Prioritize and strengthen sexual and reproductive health services including those for the prevention of—and response to—GBV, with an emphasis on mitigating barriers identified for specific vulnerable groups.
- Prioritize mental health and psychosocial support services in communities.

11.5 Concluding Remarks and Recommendations

This chapter argues that gender equality is an indispensable part of economic development and growth, and that women's role in the green economy is essential for creating resilient and inclusive economies. This chapter reviews the literature linking gender and economic development in four broad domains: (1) employment and social protection; (2) unpaid work and the care economy; (3) the green economy and gender inequality; and (4) COVID-19 as a window of opportunity for women in the green economy. Which policies can help achieve gender equality in the green industry in the post-COVID-19 recovery period? Our research suggests that if women are only provided with employment opportunities, their empowerment will only be partial. Instead, comprehensive policies—that address the educational, socio-cultural, and legal aspects of women's lives—are needed to achieve more gender equality. Without these, women's increased employment will come at the cost of their leisure time, as they are forced to shoulder unpaid childcare and household duties in addition to their newly gained salaried work. Changing "traditional" gender roles and attitudes toward women; addressing the need for childcare facilities; and ameliorating persisting inequalities in the legal system should also therefore feature high on the agenda (Nikolova and Polansky 2022). Policies which encourage women to combine childcare with work, such as shared parental leave (following the Scandinavian example) or flexible working policies, may be particularly useful. Education is an important tool for promoting gender diversity, employment, and business ownership. Many of the future opportunities for women in the green economy are likely to arise in the areas of science, technology, engineering, and mathematics. Women are still under-represented in these subjects in secondary and tertiary education and tapping this unexplored potential will be essential (Nikolova 2017). Equal access to specialized education will also ensure that women are able to utilize their skills across a variety of sectors. Cultural biases which devalue women's role in society and the economy must also be changed. Attitudes will not change overnight, since cultural formation is a long-term process, but key stakeholders (including governments, civil society organizations, private sector companies, and entrepreneurial associations) have a responsibility to raise awareness, offer training programs and networking opportunities, and promote female role models. Lastly, legal systems will require reform to ensure that equal work is equally paid and that gender-discrimination does not influence hiring (Nikolova and Polansky 2022).

References

Adamy, Joann. 2016, February 8. Companies where more Women Lead are more profitable, a new report says. *The Wall Street Journal*. https://blogs.wsj.com/economics/2016/02/08/companies-where-more-women-lead-are-more-profitable-a-new-report-says/. Accessed 15 Nov 2022

Addati, L., Cassirer, N., Gilchrist, K., and Chen, M.A. 2018. *Care work and care jobs for the future of decent work*. International Labour Organization. https://www.ilo.org/global/publications/books/WCMS_633135/lang%2D%2Den/index.htm.

Ahonen, E.Q., V. Porthé, M.L. Vázquez, A.M. García, M.J. López-Jacob, C. Ruiz-Frutos, E. Ronda-Pérez, J. Benach, F.G. Benavides, and ITSAL Project. 2009. A qualitative study about immigrant Workers' perceptions of their working conditions in Spain. *Journal of Epidemiology and Community Health* 63 (11): 936–942.

Alonso, Cristian, Mariya Brussevich, Era Dabla-Norris, Yuko Kinoshita, and Kalpana Kochhar. 2019. *Reducing and redistributing unpaid work: stronger policies to support gender equality*. Washington: International Monetary Fund.

Brainerd, Elizabeth. 2013. Missing Women in the former Soviet Union? Son preference and Children's health in the transition from communism. Unpublished manuscript, Brandeis University, Waltham, MA. https://pdfs.semanticscholar.org/03de/394a40b622d10eb005c110b2f69de12a809b.pdf. Accessed 10 Nov 2021

Brussevich, Mariya, Era Dabla-Norris, Christine Kamunge, Pooja Karnane, Salma Khalid, and Kalpana Kochhar. 2018. Gender, technology, and the future of work. International Monetary Fund Staff Discussion Note.

CARE. 2020. Rapid gender analysis during COVID-19 pandemic: Mekong sub-regional report. ReliefWeb. https://reliefweb.int/report/cambodia/rapid-gender-analysis-during-covid-19-pandemic-mekong-sub-regional-report-cambodia

Chopra, Deepta. 2015. Balancing paid work and unpaid care work to achieve Women's economic empowerment. In *IDS policy briefing 83*. Brighton: Institute of Development Studies. https://opendocs.ids.ac.uk/opendocs/handle/20.500.12413/5623.

Corossacz, Ribeiro V. 2018. *White middle-class men in Rio De Janeiro: The making of a dominant subject*. Lanham, MD: Lexington Books.

DCED. 2019. *Gender responsive green growth: a guidance note*. https://www.enterprise-development.org/wp-content/uploads/DCED-Guidance-Sheet-Gender-responsive-green-growth-introduction.pdf. Accessed 10 Nov 2021

Dey, Soma, and Palash Basak. 2017. Out of the shadows: Women and wage struggle in the RMG industry of Bangladesh. *Asian Journal of Women's Studies* 23 (2): 163–182.

Duflo, Esther. 2012. Women empowerment and economic development. *Journal of Economic Literature* 50 (4): 1051–1079.

Elborgh-Woytek, Katrin, Monique Newiak, Kalpana Kochhar, Stefania Fabrizio, Kangni Kpodar, Philippe Wingender, Benedict J. Clements, and Gerd Schwartz. 2013. *Women, work, and the economy: Macroeconomic gains from gender equity*. Washington, DC: International Monetary Fund.

Elson, Diane, and Ruth Pearson. 1981. 'Nimble fingers make cheap workers': An analysis of Women's employment in third world export manufacturing. *Feminist Review* 7 (1): 87–107.

Folbre, Nancy. 2018. Developing care: Recent research on the care economy and economic development. International Development Research Centre. https://idl-bnc-idrc.dspacedirect.org/bitstream/handle/10625/57142/IDL-57142.pdf?sequence=2&isAllowed=y.

GGGI. 2021. *GGGI gender equality and social inclusion strategy 2021-2025, leaving no one behind in the transformation towards a low-carbon, resilient world of strong, inclusive, and sustainable growth.* n.d. Available at: https://gggi.org/site/assets/uploads/2021/02/GGGI_Gender-Strategy-2021-2025-FINAL-January-1-2020.pdf. Accessed 10 Nov 2021.

Gonçalves, Terezinha. 2010. Crossroads of empowerment: The Organization of Women Domestic Workers in Brazil. *IDS Bulletin* 41 (2): 62–69.

Guo, Zhen, Monica Das Gupta, and Shuzhuo Li. 2016. 'Missing girls' in China and India: Trends and policy challenges. *Asian Population Studies* 12 (2): 135–155.

Ilkkaracan, Ipek. 2018. Promoting Women's economic empowerment recognizing and investing in the care economy. Issue paper. New York, NY: UN Women.

Inter-Parliamentary Union. 2021. Women in Politics 2021. https://www.ipu.org/resources/publications/reports/2022-03/women-in-parliament-in-2021

Jadhav, Yadav. 2009. Empowering rural Women workers through green jobs: Learning from Women's participation in the Indian forestry sector. Case study for sustain labor project on green jobs and Women workers.

Jha, Chandan Kumar, and Sudipta Sarangi. 2018. Women and corruption: What positions must they hold to make a difference? *Journal of Economic Behavior & Organization* 151: 219–233.

Joireman, S.F. 2008. The mystery of capital formation in Sub-Saharan Africa: Women, property rights and customary law. *World Development* 36 (7): 1233–1246.

Jolly, Susie. 2011. Why is development work so straight? Heteronormativity in the international development industry. *Development in Practice* 21 (1): 18–28.

Kasliwal, Ria. 2020. Gender and the gig economy: A qualitative study of gig platforms for Women workers. *ORF issue brief* 359. Observer Research Foundation.

Klasen, Stephan. 2002. Low schooling for girls, slower growth for all? Cross-country evidence on the effect of gender inequality in education on economic development. *The World Bank Economic Review* 16 (3): 345–373.

Kochhar, Kalpana, Sonali Jain-Chandra, and Monique Newiak. 2017. *Women, work, and economic growth: Levelling the playing field.* Washington, DC: International Monetary Fund.

Lagarde, Christine, and Jonathan D. Ostry. 2018. The macroeconomic gains from gender equality. VoxEU.org, December 5. https://voxeu.org/article/macroeconomic-benefits-gender-diversity.

Lahoti, Rahul, and Hema Swaminathan. 2016. Economic development and Women's labor force participation in India. *Feminist Economics* 22 (2): 168–195.

Majumdar, Ashima and Saundarjya Borbora. 2013, October 19. Social security system and the informal sector in India: A review. *Economic and Political Weekly* 48(42): 69–72. http://www.jstor.org/stable/23528577.

McKinsey Global Institute. 2015. *How advancing Women's equality can add $12 trillion to global growth.* Shanghai: McKinsey Global Institute.

McLean, Catriona. 2019. Women: powerful agents of change for an economic transition. https://www.greeneconomycoalition.org/news-and-resources/women-powerful-agents-of-change-for-an-economic-transition. Accessed 25 Sept 2021

———. 2020. The green economy mustn't forget gender. https://www.greeneconomycoalition.org/news-and-resources/the-green-economy-mustnt-forget-gender. Accessed 25 Sept 2021

Milkman, Ruth, Luke Elliott-Negri, Kathleen Griesbach, and Adam Reich. 2021, May. Gender, class, and the gig economy: The case of platform-based food delivery. *Critical Sociology* 47 (3): 357–372.

Miller, G. 2008. Women's suffrage, political responsiveness, and child survival in American history. *Quarterly Journal of Economics* 123 (3): 1287–1327.

Murphy, Clíodhna. 2013. "The Enduring Vulnerability of Migrant Domestic Workers in Europe." International and Comparative Law Quarterly 62, no. 3: 599–627. https://doi.org/10.1017/s0020589313000195.

Nelson, Julie A. 2016, January. Husbandry: A (feminist) reclamation of masculine responsibility for care. *Cambridge Journal of Economics* 40 (1): 1–15.

Neri, Marcelo Cortes. 2010, December. Good news for the poor. *The Brazilian Economy* 2 (10): 24–27.

Nikolay Angelov, Per Johansson, and Erica Lindahl. 2016. Parenthood and the gender gap in pay. *Journal of Labor Economics* 34 (3): 545–579.

Nikolova, Elena. 2017. Can diversity encourage entrepreneurship in transition economies? *IZA World of Labor* 313: 1–10.

Nikolova, Elena, and Jakub Polansky. 2022. Children and female employment in Mongolia. *Economic Systems* 46 (3): 101019.

Nizalova, Olena Y., Tamara Sliusarenko, and Solomiya Shpak. 2016, February. The motherhood wage penalty in times of transition. *Journal of Comparative Economics* 44 (1): 56–75.

OECD. 2020. *OECD Policy responses to coronavirus (COVID-19): making the green recovery work for jobs, income and growth.* https://www.oecd.org/coronavirus/policy-responses/making-the-green-recovery-work-for-jobs-income-and-growth-a505f3e7/. Accessed 25 Sept 2021

Olivetti, Claudia, and Barbara Petrongolo. 2016. The evolution of gender gaps in industrialized countries. *Annual Review of Economics* 8: 405–434.

Ostry, David J., Jorge Alvarez, Raphael A. Espinoza, and Chris Papageorgiou. 2018. *Economic gains from gender inclusion: New mechanisms, new evidence.* Washington DC: International Monetary Fund.

Puri, Lakshmi. Speech by Lakshmi Puri on economic costs of violence against Women. UN Women, 14 September 2016. https://www.unwomen.org/en/news/stories/2016/9/speech-by-lakshmi-puri-on-economic-costs-of-violence-against-women.

Ridgeway, Cecilia L., and Shelley J. Correll. 2004. Unpacking the gender system: A theoretical perspective on gender beliefs and social relations. *Gender and Society* 18 (4): 510–531. http://www.jstor.org/stable/4149448.

Rose, Sonya O. 1986. 'Gender at work': sex, class and industrial capitalism. *History Workshop Journal* 21 (1): 113–132. Oxford University Press.

Ross, Michael L. 2008. Oil, Islam, and women. *American Political Science Review* 102 (1): 107–123.

Salardi, Paola. 2016. The evolution of gender and racial occupational segregation across formal and non-formal labour Markets in Brazil, 1987 to 2006. *Review of Income and Wealth* 62 (1): 68–72. 83-86.

Schulz, Anna. 2021. Guest article: Green growth is at the heart of efforts to achieve gender equality and Women's empowerment while delivering on the SDGs and the Paris agreement: SDG knowledge hub: IISD. *SDG Knowledge Hub*, September 26, 2021. https://sdg.iisd.org/commentary/guest-articles/green-growth-is-at-the-heart-of-efforts-to-achieve-gender-equality-and-womens-empowerment-while-delivering-on-the-sdgs-and-the-paris-agreement/. Accessed 26 Sept 2021.

Siddiqi, Dina M. 2009. Do Bangladeshi factory workers need saving? Sisterhood in the post-sweatshop era. *Feminist Review* 91 (1): 154–174.

UN Women. 2014. *The world survey on the role of Women in development 2014: Gender equality and sustainable development.* New York: United Nations.

———. 2018. Promoting Women's economic empowerment: Recognizing and investing in the care economy. https://docs.euromedwomen.foundation/files/ermwf-documents/8098_4.193.promotingwomen percentE2 percent80 percent99seconomicempowerment-recognizingandinvestinginthecareeconomy.pdf.

UNEP. 2020. *Resilient and inclusive circular economy to build Back better and greener.* Policy brief. Available at: https://wedocs.unep.org/bitstream/handle/20.500.11822/32922/Circularpercent20Economy_LAC.pdf?sequence=1&isAllowed=y. Accessed 27 Sept 2021.

UNEP and UN Women. 2020. Gender and climate change in the context of COVID-19. Think piece. Available at: https://www.preventionweb.net/files/73854_empowerongenderandclimateinthecovid.pdf. Accessed 26 Sept 2021.

UNICEF. 2021. *Social Protection and gender equality: Outcomes across the lifecycle.* Full Report. https://www.unicef.org/media/112591/file/Social-Protection-Gender-Equality-Outcomes-Across-the-Lifecycle-Full-Report.pdf. Accessed 28 April 2023

UNIDO. 2021. *Policy assessment for the economic empowerment of Women in green industry.* Cambodia report. Available at: https://www.unido.org/sites/default/files/files/2021-06/Cambodia_Countrypercent20Report_Final_ENG_0.pdf. Accessed 27 Sept 2021.

Williams, Fiona. 2010. Migration and care: Themes, concepts and challenges. *Social Policy and Society* 9 (3): 385–396.

Wilson, Kalpana. 2015. Towards a radical re-appropriation: Gender, development and neoliberal feminism. *Development and Change* 46 (4): 803–832.

Young, Marisa C. 2010. Gender differences in precarious work settings. *Relations Industrielles/Industrial Relations* 65 (1): 74–97. http://www.jstor.org/stable/23078260.

CHAPTER 12

Disaggregating Data by Sex and Gender to Enable Inclusive Digital and Green Transitions in Ethiopia, India, and the USA

Kristie Drucza and Natalia Pastori Curbelo

12.1 Introduction

It is a truism that growth reduces gender inequalities, yet we only need to look to Japan to see that there can be national growth without gender equality. Similarly, the idea that the environment is more connected to women ("mother nature") or that traditional indigenous cultures are more environmentally friendly are again not necessarily true (just consider "back burning" and "smoking" practices). Other myths—such as men are better leaders or more technologically skilled or that low-income countries are more climate-friendly (just Google the "Top 10 Countries with the Worst Air Pollution")—harm mainstream debates around the

K. Drucza
Includovate, Kampala, Uganda
e-mail: kristie.drucza@includovate.com

N. P. Curbelo (✉)
Includovate, Queanbeyan East, NSW, Australia
e-mail: natalia.curbelo@includovate.com

K. Drucza et al. (eds.), *Centering Gender in the Era of Digital and Green Transition*,
https://doi.org/10.1007/978-3-031-38211-6_12

environmental and gender issues. Myths, ideologies, and generalizations have indeed confused our evidence-base and impact our ability to properly address and solve these global problems.

The world's population is more climate-aware than ever before. More than 70 percent of global GDP is now covered by a net-zero target, with many advanced economies aiming to decarbonize by 2050 (Pennington 2022). Digital and green transitions offer a way to integrate sustainable growth, but can only do so if they include mechanisms to address social inequalities. By not presenting data that is sex-disaggregated and gender-sensitive, green and digital transitions are likely to further increase inequalities. The Organization for Economic Co-operation and Development (OECD) makes this clear, arguing that "in a green transition, many of the people who could be hit the hardest are lower-income demographics who still rely on cheap fossil fuel-powered resources rather than being financially able to invest in more expensive green technologies" (OECD 2021). COVID-19 accelerated digital transitions, with more people forced to work from home, yet gender inequality increased during the same period.[1] Using the feminist standpoint theory and the concept of intersectionality, this chapter explains the way different vectors and identities combine to create unique disadvantages and deeper pockets of exclusion. This chapter examines three different countries, chosen to represent different income levels, demographics, and socio-political contexts: Ethiopia (low-income), India (middle-income), and USA (high-income). Comparisons between these three very different countries can help us analyze trends in access and consider how women use, benefit, and are excluded from digital and green technologies.

This chapter uses data on mobile phone access, internet subscriptions, and environmental science degrees as proxy indicators for digital transitions. It also uses the Green Industrial Performance (GIP) index, as well as air pollution and greenhouse gas emissions (GHG) as green transitioning indicators. However, since the latter are not sex-disaggregated, they are compared with national gender and equality indicators from the World Economic Forum's (WEF's) *Gender Gap* report of 2022 and the Gini coefficient. Our data reveals, in summary, that without concerted effort, digital and green inequalities will not simply reduce at the same rate that wealth is created, even if growth is environmentally sustainable. This chapter concludes by mapping a path toward more equal access to green and digital opportunities (current and emerging) via improved sex-disaggregated and gender-sensitive data.

[1] The World Economic Forum's *Global Gender Gap* report (2021) showed that closing the global gender gap would take an additional 36 years as a result of the impact of the pandemic.

12.2 Feminist Standpoint Theory and Intersectionality

Researchers have used feminist standpoint theory and intersectionality to examine and understand gender in the era of green and digital transitions. Feminist standpoint theory (under the purview of the social sciences) argues that research, knowledge dissemination, and scientific discourse must consider the perspectives of women, including those who have been marginalized. The term "standpoint" indicates that the theory goes beyond merely capturing women's perspectives to designate moral and political authority to women (Gurung 2020). In the context of green and digital transitions, this allows a deeper analysis of existing norms, policies, and trends as they relate to gender in agriculture, technology, and education. The feminist standpoint theory is not new. It has been evident in the writings of feminist scholars such as Smith, Harding, Hartsock, and Haraway since the early 1970s and 1980s (Halpern 2019) and derives its principles from feminist theories dating back to Marx and Hegel. The basic principle of the theory is that all knowledge is situated, interpreted, and localized (Halpern 2019), and that marginalized groups are situated in ways that allow them to see more than those who are not marginalized (Halpern 2019). Standpoint theorists argue that oppressed people have different knowledge, experiences, and perspectives to the mainstream. Historically, institutionalized scientific knowledge has not been produced by women or other oppressed groups but created by men (*Black Feminisms*, 2021). The framing of "knowledge" is not neutral, therefore, but repeats and reinforces gender norms. Scientific knowledge that has been produced and institutionalized by men cannot effectively inform the lives of women and other oppressed groups.

This chapter also looks at data through the lens of intersectional feminism (or "intersectionality"). First coined in 1989 by Kimberlé Crenshaw, an American law professor, intersectionality is a "prism for seeing the way in which various forms of inequality often operate together and exacerbate each other" (UN Women 2020). Intersectionality emphasizes that social identities can overlap and interact, creating compounding impacts and experiences of discrimination, oppression, and domination. Crenshaw argues that intersectionality requires us to study the power structures that impact marginalized individuals and groups, specifically black women (Adewunmi 2014). An intersectionality perspective is now applied across socio-political, academic, and legal discourses all around the world. Adding such a perspective to feminism allows gender equality advocates to be more inclusive of other groups that have also been historically marginalized—such as women of color, the LGBTIQA+ community, lower castes and classes, certain religious groups, and those from disadvantaged socio-economic backgrounds.

This chapter has been written by women—one with a disability and three others from the Global South—who use English as a second or third language; a feminist standpoint theory and intersectionality frame our findings and analysis.

12.3 Sex-Disaggregated and Gender-Sensitive Data and Analysis

While the terms "gender-disaggregated data" and "sex-disaggregated data" are often used interchangeably, their meanings are different. Sex-disaggregated data is collected and presented in two distinct and separate categories, one that considers men and one that considers women. When data is disaggregated by sex it is expected to "reflect differences and inequalities in the situation of women and men in all areas of life" (UN 2016). However, most sex-disaggregated data still use the outdated binary notion of "men" and "women," and exclude the third and fourth genders, as well as those who are transitioning. Despite this, most studies do not include sex options other than "male" or "female." Even if they did, few respondents—especially in low-income countries where heteronormativity is expected—would be likely to tick any other box but "male" or "female." Consequently, it is difficult to find fully gender-disaggregated data.

In order for research to benefit all genders, it is important that data is disaggregated by sex from the start of the research process. A gendered analysis of the data should follow, to properly examine intersections between gender, age, and other key factors such as education, health, marital status. Gender—as opposed to sex—is not a static concept. It is shaped by socially constructed roles, expressions, identities, and behaviors. Gender-sensitive data integrates a gender perspective in the collection, analysis, and presentation of data (EIGE 2019) and has the capacity to reveal differences that may be the result of gender norms and social expectations. Such factors are important when trying to address social inequalities and provide solutions for sustainable change. Gender-sensitive data should have the following characteristics:

- It should reflect gender-based issues
- It should be based on concepts that adequately depict the diversity of men and women's lives
- It should consider—in its data collection methodology—stereotypes and sociocultural factors that can lead to gender biases (Doss and Kieran 2013).

A gender analysis can help illustrate the differences between men and women in terms of their different needs, roles, statuses, priorities, capacities, constraints, and opportunities. It can usefully examine the evolving roles and relationships between men and women and analyze how they interact in various contexts. A common error in approaching gender analysis is to adopt a heteronormative approach and to only study women's positionality. By excluding the gendered experiences of men and non-binary people, we limit our understanding of gender relations (Cameron and Stinson 2019). Moreover, women are not a homogenous category, and a good gender analysis always incorporates the multiple intersecting identities of different groups of women (see Box 12.1).

Box 12.1 An Example of Gender-Sensitive Analysis
Environmental conditions affect the lives of men and women differently and unequally, although they are often portrayed as affecting everyone in the same way. For instance, during the 2004 tsunami, 77 percent of those who died in four villages in Indonesia's North Aceh were women (Singh et al. 2010). Many commentators found it surprising that more women than men were killed, but in these villages, men do the fishing and were out in boats when the tsunami hit. Missing from the 77 percent statistic, therefore, is the "why" that a good gender analysis can answer.

Further analysis of the Aceh situation, using an intersectional lens, shows that—in certain locations—Muslim women living next to the ocean were more affected than men (as well as other groups of women). In these areas, women are not taught to swim. They also wear burkas (which are difficult to swim in), wash and dry clothes on water banks, and do weaving and other chores under the shade of palm trees on the beach. Cultural and religious norms, therefore, shaped gender roles, responsibilities, and clothing in North Aceh and resulted in more women dying there during the tsunami.

Men, women, and other genders are not treated the same by society, and when a disaster occurs inequities can magnify. By extension, inequalities in other areas (e.g. education, health, technology, or household decision-making) can result in one gender being more affected by environmental disasters and climate changes than others (UN 2016).

12.4 Digital and Green Transitions

The following sections present sex-disaggregated data for digital and green transitions in three countries—Ethiopia (low-income), India (middle-income), and the USA (high-income). Using a gender data analysis, it addresses inequalities relating to the digital and green transitions in these countries and emphasizes the differences between each country's resources and capacities to address these issues.

Green Transitions

A green transformation requires economies to transition away from fossil fuels to energies that are "circular, sustainable and decarbonized" (*The Beautiful Truth*, 2022). A vital pillar of a green transition is figuring out how to get the most out of materials by improving design, recycling, and repurposing in order to eliminate waste. There is a great need for equitable green transitions in the key sectors (i.e. energy, transport, and food systems) that produce the majority (>90 percent) of GHG emissions (WB Live 2021). Ensuring an inclusive, fair, and just transition will be central to enabling an equitable green transition and ensuring that those who can least afford it are not left further behind.

Table 12.1 compares three national green transition indicators: the GIP index (De Alba and Todorov 2022), air pollution, and GHG emissions. Since these indicators are not disaggregated by sex, additional information is added from national gender, and equality indicators, specifically from the World Economic Forum's *Gender Gap* report of 2022 and the Gini coefficient.[2]

The USA is the richest of the three countries, has the greenest industries and the lowest air pollution, but is polluting the world with the highest greenhouse gas emissions. India has the worst air pollution (5th in the world), mid-level greenhouse gas emissions, and needs to work on green industrial performance. Ethiopia has mid-level air pollution, hardly any greenhouse gas emissions but a terrible green industrial performance. It is also the poorest country, suggesting it will have limited resources to rectify its industrial performance (EDGAR 2021). Out of the 146 countries

[2] The Gini coefficient index measures the extent to which the distribution of income or consumption among individuals or households within an economy deviates from a perfectly equal distribution. A Gini index of 0 represents perfect equality, while an index of 100 implies perfect inequality (WB Glossary).

Table 12.1 National statistics by country

Indicator	*Ethiopia*	*India*	*USA*
Population	119 million	1.417 billion	339 million
GDP per capita (US$)	925.08	2256.59	70,248.63
GIP index rating (out of 112 countries)	100	74	13
Air pollution (μg/m)	23.9	58.1	10.3
Greenhouse gas emissions	0.30 percent (2018)	7.07 percent (2018)	12.30 percent (2018)
WEF's *global gender gap* report 2022	0.710	0.629	0.769
Gini coefficient	35 (2015)	35.7 (2019)	41.4 (2018)

included in WEF's *Global Gender Gap* report, the USA ranks 27th, Ethiopia 74th, and India 135th. However, the USA has the highest Gini coefficient, showing the greatest disparities between rich and poor households. This suggests that even if the USA does make further progress in gender equality, pro-poor policies are unlikely to gain traction there.

As this mini-synopsis reveals, ensuring an inclusive and fair green transition is complicated because wealth does not automatically lead to equality, and each country has complex and diverse challenges that require a tailored approach to solve.

12.5 Digital Transitions

Digitization concerns "the way many domains of social life are restructured around digital communication and media infrastructure" (Brennen and Kreiss 2016). Digitization can help (along with other interventions) improve effectiveness, efficiency, interoperability, and transparency; help tack climate change; and support green transitions (George et al. 2021). The Internet is not automatically good for the environment, accounting for around ten percent of global electricity demand (Lozano 2019). Nevertheless, digitization is accelerating and women are being left behind. This section presents data on women's access to mobile phones, internet subscriptions, jobs and Science, Technology, Engineering, and Mathematics (STEM) degrees and completes an analysis of the "gender digital divide" (see Tables 12.2 and 12.3).

Table 12.2 National Statistics on the Digital Transition by Country (World Bank Database)

	Ethiopia	*India*	*USA*
Mobile phone ownership	27% women 54% men	63% women 79% men	98% women 97% men
Internet subscription	20.6%[a]	15% women 25% men	93% women 94% men
Women with STEM degrees	17.3% (2010)	42.7% (2018)	34% (2016)
Women in computer science jobs	12%[b]	34 % (2017–18)	26%

[a]This figure accounts for the total population (Gilbert 2021)
[b]This figure is for the African continent. Ethiopia-specific data was not included. (*New Business Ethiopia* 2022)

Table 12.3 National Statistics on the Digital Transition in Ethiopia. (World Bank Database)

	Ethiopia
Internet users by gender and location	Rural: 0% women; 2% men Urban: 14% women; 30% men
Internet users by gender and household wealth	Poorest: 0% women; 1% men Richest: 12% women; 24% men

Ethiopia

An important indicator of digital gender gaps is women's representation in STEM research and education. In Ethiopia, women hold only 17.3 percent of the country's STEM degrees and across Africa only 12 percent of the computer science jobs (disaggregated data on STEM jobs in Ethiopia is not available). Many factors limit women's participation in STEM. Women's education in Ethiopia is strongly influenced by sociocultural norms and attitudes that restrict their enrolment in higher education. Sexual violence and biases against women also affect their academic performance and psychological well-being, reducing their participation in higher education and STEM programs (Melak and Singh 2021). Women have less access to the internet than men. A study conducted by MacQuarrie et al. (2022) found that only 14 percent of urban women have access to the Internet, and rural women do not have access at all (0 percent). Among the poorest women, internet use is similarly non-existent

(0 percent). Even if women had equal access to the internet as men, "the nature of the available ICT content was more appealing to men than to women" (Geldof 2011). Ethiopia's Information and Communications Technology (ICT) landscape "is greatly determined and controlled by state-owned monopolies ... that constitutes a major stumbling block for innovation and widespread access to ICT services" (Geldof 2011). In Ethiopia, gender norms limit access to finances, and the high cost of internet, lead to women having less access to digital services than men. Women are 50 percent less likely than men to own mobile phones. A study by LeFevre et al. (2020) found that only 27 percent of women in Ethiopia owned mobile phones, compared to 54 percent of men. LeFevre's intersectional analysis showed that women living in rural areas and poorer households are 25 percent less likely to own a phone than their wealthier urban counterparts. In general, women are more likely to own a mobile phone if they are older, unmarried, wealthier, more educated, urban, or have two or less children (LeFevre et al. 2020). A lack of official government identification also hinders women's access to—and ownership of mobile phones. Kappler (2019) finds that across Sub-Saharan Africa, women are 9 percent less likely to have official identification cards/papers, and this disparity is twice as large in Ethiopia. Sex-disaggregated data relating to internet access and STEM participation are available in Ethiopia, but are very limited, and there is little analysis of gender issues. Even with this limited data, it is clear that many Ethiopian women are likely to be left behind by digital transformations. As Geldof (2011) has argued in her article on Ethiopia and Malawi, the ways in which ICTs are used by men and women reflect the sociocultural and economic contexts in which they live, since gender and ICTs processes are both naturally fluid and dynamic.

India

India's access to technology across sectors and groups has increased over recent years. However, gender gaps continue in terms of access. The Global System for Mobile (GSM) technology's *Mobile Gender Gap* report of 2020, for example, showed that while 79 percent of men own mobile phones only 63 percent of Indian women do. Women's mobile phone ownership rates have remained relatively stable since 2017 (around 60 percent), but changing market dynamics and the increasing affordability of the internet have affected the gender gap. In 2020, only 15 percent of women had access to the internet compared to 25 percent of men (GSMA

2020)—the widest gender gap in this regard across the Asia-Pacific region (Mitali and Uppadhayay 2021). This should be a cautionary tale, for India demonstrates that increases in wealth and mobile/internet affordability, will not level gendered digital divides alone. Looking through an intersectional lens, it is clear that access gaps emerge not only as a result of disparities and inequalities (within and between families and rural/urban locations) but through gendered expectations of "decency," matrimony, and motherhood (Mitali and Uppadhayay 2021). Women of all ages, whether married or unmarried, deviate from socially constructed gender norms and expectations when they use the internet on mobile phones. Such usage is assumed to impede the capacity of mothers to care for their kids or to carry out household chores, and to lead to unmarried women pursuing "immoral" behaviors, such as romances with men. These norms and assumptions force women to choose between being seen to be "good" or being informed (Barboni et al. 2018).

Mobile phones have become a critical way of sharing information, opportunities, and services, and women's low access excludes them from new growth opportunities and information. Lack of access also limits their independence (Barboni et al. 2018) and access to health care: during the COVID-19 pandemic, for example, women were less likely to book vaccination slots than men (CEDA 2021).

In developing countries where there are limited opportunities for economic mobility, STEM careers present a clear path toward women's financial freedom and can further help improve the socio-economic status of families (Stoet and Geary 2018; Khazan 2018). Indian women currently earn 43 percent of all STEM degrees across the country—the highest county-level percentage by women globally (Lennartz 2023). Yardi et al. (2017) argue that the higher number of female STEM graduates in India is the result of parents pushing their daughters into STEM for socio-economic reasons (it is seen as a lucrative and acceptable profession for women). However, gender gaps widen considerably when we analyze employment stats, with Indian women holding only 14 percent of the nation's overall STEM jobs (Jayan 2020). This disparity is attributed to male domination of the STEM field, women's lack of mentorship, and heavier gendered responsibilities in terms of matrimony and motherhood (Mitali and Uppadhayay 2021). Fanny von Heland, a Swedish Science Counsellor, argues that this employment gap wastes women's STEM talents, as graduates leave to pursue careers in other sectors, or simply do not work at all (Jayan 2020).

USA

Compared to its under-developed and developing counterparts, the USA fares well in terms of equitable gendered access to digital technology. The Pew Research Center's (2021) research on mobile ownership indicates that 98 percent of American women and 97 percent of American men own a mobile phone, and that 85 percent of both genders own a smartphone. Internet access is also more equitable with 15 percent of women relying solely on smartphones for access to the internet (without home broadband) as opposed to 14 percent of men. However, from an intersectionality perspective, equitable access to mobile technologies may differ between different groups of women depending on their income, inter-generational wealth, age, race, place of residence, and disability, and so on. For example, in 2021, 94 percent of men and 93 percent of women in the USA used the Internet, but only 75 percent of people over the age of 65 did. In addition, only 44 percent of households with incomes below US$30,000 per annum have a home broadband connection (PEW 2021). Despite the overall equitable access to mobile phones and the internet, the majority of those participating in the online economy are men (Farrell and Greig 2017). Women are less likely to take advantage of opportunities for generating digital income, even when they have access to the internet and mobile phones. American women's representation in STEM degrees compared to jobs is another indicator of the gender technology gap. Although women are proportionally more likely to have STEM degrees across all racial groups and at nearly all degree levels (53 percent at bachelor's, 60 percent at master's, 48 percent at research doctorate, and 58 percent at professional doctorate, see Fry et al. 2021), women hold only 27 percent of the country's STEM jobs (Census Bureau 2021). Sex-disaggregated data from the USA reveals that access to mobile phones and the internet may not necessarily enable women's social and economic empowerment. Other factors besides access can contribute to disadvantages (OECD 2018). A study by Yardi et al. (2017) comparing India's better record of attracting women into STEM jobs (compared to America), found that although the USA performed better overall in terms of overall gender equality, American women faced some gender bias and discrimination at school, which their Indian counterparts did not report. Factors contributing to gender disparities in STEM jobs in the USA include sexism, male domination of the STEM industry, and women's heavier gendered responsibilities in terms of matrimony and motherhood (which are not equally shared by their male spouses) (Yardi et al. 2017).

12.6 Looking to the Future: The Economy and Automation of Green Transition

As shown previously, gender inequalities persist across many sectors and areas and can result in women being disproportionately left out of digital/green transitions and being more affected by environmental disasters and climate change. These vulnerabilities may be further compounded by automation, with the World Economic Forum reporting that 44 percent of jobs are vulnerable to future automation. Additional factors also need to be considered when assessing women's access to opportunities arising from any transformation, especially automation, such as women's representation in different sectors and unpaid labor at home (OECD 2018). This section builds upon the available green and digital data to discuss the role of women in the three national economies and how automation may help or hinder their progress.

Ethiopia's economy is largely based on agriculture, with low industrialization. The 2013 Time Use Survey showed that women are three times as likely as men to spend time on unpaid household activities such as collecting firewood or water (IMF 2018). Women in rural areas also work more—between 13 and 17 hours a day across the country, or two to three times the amount of time worked by men (Mulema and Damtew 2016). Automation in water and fuel is very promising for women and would free up their time for other pursuits including self-care, income generation, and leisure activities.

Women currently have fewer employment opportunities than men. In Ethiopia, only 74.2 percent of women participated in the labor market compared to 86.5 percent of men. Women are also more likely to work informally. For instance, 58 percent of women working in small-scale manufacturing jobs are unpaid family workers, while 40 percent of unpaid workers are male. These numbers are similar in the agricultural sector (IDR 2020). Women's participation in the formal labor market through manufacturing and retail is increasing (particularly in the textile/garment, food/beverage, and agro-industrial processing/export sectors). Although the Ethiopian government is developing industrial policies and promoting growth in the manufacturing sector (where women are frequently employed), most of these jobs are likely to become automated over the next 10 years (Gaus and Hoxtell 2019). A report developed by Citi Group and the Oxford Martin School (2016), for instance, estimates that in the next decade or two, 85 percent of Ethiopia's jobs will become automated. Given how many women are precariously employed today, women are likely to fall further behind men, with these challenges piling on top of others (including less access to STEM jobs and more time dedicated to unpaid labor).

India

The Time Use Survey indicates that across both urban and rural India, women on average spend 433 minutes a day on unpaid caregiving/household chores a day compared to the 173 minutes spent by men (Ministry of Statistics & Programme Implementation 2019). This has huge impact; the gender norms that burden women with the majority of household responsibilities affect women's work-life balance, mental well-being, and representation in professional spaces (Chouhan 2020).

The Periodic Labour Force Survey (PLFS 2019–2020) finds that Indian women's overall participation in the labor market is significantly lower than men's. Disparities are uneven across sectors and locations. For instance, in rural areas, women and men are equally represented in the manufacturing sector (7.3 percent), but disparities are much higher in construction (5.6 percent of women compared to 15 percent of men), and in the rural trade, hotel, and restaurant sectors (3.7 percent of women compared to 9.2 percent of men). More Indian women also work in the rural agricultural sector than men (75.7 percent and 55.4 percent respectively). In urban settings, there are less disparities in manufacturing (22.4 percent of women compared to 20.3 percent of men) as well as in the trade, hotel, and restaurant sectors (22.3 percent of women compared to 28.9 percent of men). Urban women are least likely to be employed in construction and transport (4.9 percent compared to 12 percent of men) and in the storage and communication (3.6 percent compared to 12 percent of men) (Bhardwaj 2022).

In order to generate income and own new technologies, increasing numbers of rural Indian women work informally as agricultural laborers (Satyavathi et al. 2010). However, the majority of agriculture-related technological research does not "see" women as farmers, assuming that women's primary roles and responsibilities are related to matrimony and motherhood. This means that research outputs and recommendations are gender-biased and fail to reflect the reality of women's work, both paid and unpaid (Jiggins 1986). There is a need, therefore, to properly "see" women in agriculture-related technological research and to properly understand their needs, priorities, capacities, and concerns, so that green transitions can progress gender equality rather than hindering it.

Although "men's jobs" may be more at risk of automation than women's, overall women are more excluded from the workforce. Low-skilled and unskilled jobs have a higher potential for automation, especially in the following sectors: construction, manufacturing, transport, storage, communication, financial, business, and legal services, as well as trade, hotel, and restaurants (Singh et al. 2019).

USA

The US Bureau of Labor Statistics (BLS 2021) reports that women made up the majority of the workforce in several sectors in 2019, including education and health (74.8 percent), financial activities (52.6 percent), and leisure and hospitality (51.2 percent). During COVID-19 lockdowns, however, women left the workforce in large numbers, perhaps because the burden of childcare fell disproportionally on them, with women spending 7.1 hours daily caring for children below 13 years, as opposed to the 4.9 hours spent by men (BLS 2021; Time Use Studies). Unless digital transitions address affordable childcare, women will be left behind, regardless of their access to the internet and digital technologies.

While automation can increase production, it is also likely to decrease employment in sectors like manufacturing, which the BLS (2022) estimates will contract annually by 7.4 percent from 2021 to 2031 (BLS 2022). In the case of the USA, automation is most likely to affect the largest occupational categories in the economy, that is, office, food service, production work, and customer service and retail sales (McKinsey Global Institute 2019). Other industries anticipating rapid declines in human employment over the same decade (2021–2031) include newspaper/periodical/book and directory publishers within the information sector (expecting contractions of 2.7 percent annually as a result of competition from digital media). There will be less need for secretaries and administrative assistants, with improved software tools, communication technologies, and AI-based virtual assistants taking over such tasks. The move to automation requires further monitoring to determine the net effect by gender.

12.7 The Need for Action

The data from COVID-19 as well as from India's Green Revolution demonstrates why disaggregating and analyzing intersectional data is necessary for equality to prevail. Most research on technology does not consider women's gendered positionality within society, or—if it does—it stereotypes their disproportionate unpaid responsibilities by presuming it is their only role and vocation. This means that research outputs and recommendations are gender-biased and fail to reflect women's lived realities as paid (and unpaid) workers (Jiggins 1986). As a consequence, it cannot be assumed that inequities will automatically reduce with growth or progress.

India's Green Revolution of the 1970s brought mechanization that pushed the poorest rural women out of employment and left rural women even further behind. In the majority of cases, men were given training to help run the machines, while women were excluded from training processes as a result of gender norms relating to unpaid labor, unequal decision-making power, and technological access (Satyavathi et al. 2010). The current green and digital transitions are showing similar signs of gender inequality, with women comprising only 22 percent of AI professionals globally (New Business Ethiopia 2022). The World Web Foundation's Gender Gap Audit (2016) also stated that "without a major escalation of policy effort and investment, most of the benefits of technological change[3] ... will be captured by men—making gender inequality worse, not better." The lack of women involved in creating technologies will again result in gender-biased solutions.

COVID-19 also set back equality. In Ethiopia, 64 percent of the workers laid off in April 2020 were women, despite women only making up 42 percent of the workforce (Wieser et al. 2020). Moreover, 25 percent of male-headed households reported diminished or complete loss of incomes, compared to 44 percent of female-headed households. Women's coping strategies were also different to men's—Ethiopian women reported reducing food consumption to a greater extent than their male counterparts who bridged financial gaps by drawing on their savings (Wieser et al. 2020). Women tend to absorb household and social shocks—both of which will increase with climate change—more than men. The longer-term effects of these coping strategies during the COVID-19 pandemic, and a deeper exploration of risk responses, are needed.

[3] The study only covered the following African countries: Kenya, Mozambique, Ghana, Uganda, Nigeria, and Egypt.

COVID-19 also exposed entrenched discriminatory attitudes. A study by Dalberg in 2021, for example, showed that women from low-income households in India were disproportionately impacted by the COVID-19 crisis, particularly historically vulnerable women, such as Muslim, migrant, single, separated, widowed, or divorced women. Desai et al. (2021) also found evidence of stigmatization, with the Muslim community being blamed for spreading the disease and Muslim-owned stores being boycotted, with resultant losses of employment and income for the Muslim community. Women made up just 24 percent of the workforce prior to the pandemic but accounted for 28 percent of all job losses during the crisis (Dalberg 2021). Women's unpaid workload increased during COVID lockdowns, as schools were closed along with shops and other services, making it more difficult for them to remain in the workforce. These difficulties were exacerbated by discriminatory attitudes—with Indian men's chances of being employed during the lockdown 0.1 percent compared to the pre-lockdown period, while for women it was 0.06 percent (Desai et al. 2021).

Women in the USA were also more severely impacted by the COVID-19 pandemic than men. Bateman and Ross (2020) reported that before the COVID-19 crisis, nearly half of all working women (28 million) were in low-wage jobs, with a higher proportion of black (54 percent), Hispanic or Latina women (64 percent) in low-paid jobs compared to white women (40 percent). The restrictions imposed during the pandemic resulted in unprecedented unemployment. Women were particularly affected due to working mainly in low-wage and face-to-face jobs such as retail sales and hospitality that could not easily transition from in-person to remote work. Clapp (2023) showed that women were more likely to lose their jobs at the beginning of the pandemic and that their recovery was slower than men's. In essence, the market is a microcosm of social inequities. Across all three countries, COVID-19 starkly exposed how market and employment discrimination inflame inequities. Violence against women also increased during the COVID-19 lockdowns, along with teenage pregnancy (UNDP 2020). According to the WEF Gender Gap Report of 2021, the global gender gap widened as a result of COVID; prior to COVID, their estimates were that gender equality would be achieved in 99.5 years; after COVID that figure escalated to 135.6 years. Historical data, such as from India's Green Revolution and the 2016 Web Foundation's Gender Gap Audit, also suggests that inequalities will not be adequately reduced merely through digital and green transitions.

A Pathway Toward Equality

Sex-disaggregated data provides the foundation for policies and strategies that can support gender equality in the digital and green transitions. Gender analyses are also needed to pinpoint which variables are likely to increase inequality and which offer hope for course correction. Using sex-disaggregated and gender-analyzed data, and comparing country-level lessons, the following section helps map a path toward more equal access to the green and digital opportunities (current and emerging) in each country. Data reveals that Ethiopian women who are older, unmarried, wealthier, more educated, residents of urban areas, and have two or fewer children are more likely to benefit from the green and digital transitions. These intersectional issues should be factored into any policy solution, especially considering that automating the labor force is likely to have a disproportionately negative impact on women's employment. However, it is also possible that women can advance with an equitable green and digital transformation. Ensuring rural women have access to information and other services like irrigation, household water, and electricity will help reduce their domestic burden, and in turn, free up their time for other pursuits (Mulema and Damtew 2016).

Ethiopia can learn from India that—even with increased wealth and more affordable digital access—the digital gender gap will not reduce fast enough without additional efforts. More specifically, if all the women in Ethiopia's parliament can push for women's increased digital access and STEM enrolment, it is possible for Ethiopia to do better than India at the same stage of growth in closing the gender digital gap. Affordable green water and fuel solutions should be central to this transformation. Women are more likely to prioritize these technological innovations than men, because they provide focused solutions for women's needs.

Women in India have greater access to digital technologies and higher education, including STEM degrees, than in Ethiopia, but women are more absent from the workforce. In general, a gender access gap emerges in India due to intra-family and patriarchal gendered expectations that assign domestic chores to women more than to men. The promise of automation, and the benefits of any green and digital transition, may not equitably benefit women, especially in rural areas where women are disproportionately excluded from the workforce. To ensure these trends are corrected, universities should have mentorship programs for women enrolled in STEM, as well as bridging programs to help women secure jobs in STEM after they have completed their studies. Meanwhile, STEM-related employers should reform (or add) diversity, equity, and inclusion (DEI) policies to their companies and businesses. Such policies can create a more women-friendly culture, by promoting flexible arrangements that allow women to continue working after marriage and childbirth (policies that would have helped many women retain their jobs during and after COVID-19). A behavior change campaign around men's domestic responsibilities is also advised.

In the USA, key intersectional factors (income, inter-generational wealth, age, race, place of residence, and disability) need to be considered in any green and digital policy. Other factors of concern are the male domination of certain industries, gender bias or discrimination at school, sexism, and women's disproportionate unpaid domestic duties. Moreover, women, especially racial minorities, should be encouraged through seed funding to engage in digital income generation opportunities. As in India, there is a need to promote DEI policies in STEM organizations to entice more women to enter and stay in STEM jobs. Regular conversations around gender roles and their impact on women's progression across professions are needed to shift sexist attitudes so that women can enjoy work and study without harassment. Any digital transition that does not come with affordable childcare will leave women further behind in the USA, regardless of their access to the internet and digital technologies.

12.8 Concluding Remarks and Recommendations

In this chapter, we used a feminist standpoint theory and the concept of intersectionality to frame the current data and indicators that relate to digital and green transitions and explore how different vectors and identities combine to create unique disadvantages within and across countries. It shows that across three countries—Ethiopia (low-income), India (middle-income), and the USA (high-income)—"green" data is less likely to be sex-disaggregated than digital data. If the organizations that influence policy do not collect sex-disaggregated data or do not analyze that data by gender and other intersectional categories, then policies will not change. Careful steering and gender assessments are needed to ensure that women can equitably benefit from any transition or change. We need sex-disaggregated data, but we also need a gender analysis of that data. A good gender analysis always incorporates intersectionality. This level of granularity can confuse policymakers who prefer simple solutions and numbers.

Artificial intelligence (AI) can overcome our human bias toward simplicity, but likely only if those involved in gender analyses and programming AI have diverse backgrounds. Ensuring an inclusive and fair green transition is complicated and will require a tailored approach per country and per identity vector, along with close monitoring and a gender analysis of any data generated.

This chapter shows how a gender analysis of existing data can be used to inform policymakers and prioritize areas for future attention and resource allocation—that, in turn, can steer a course toward a fair, green, and equal future.

References

Adewunmi, B. 2014. Kimberlé Crenshaw on intersectionality: "I wanted to come up with an everyday metaphor that anyone could use". *The New Statesman.*

Barboni, G. et al. 2018. A tough call: understanding barriers to and impacts of women's mobile phone adoption in India. Harvard Kennedy School. Evidence for Policy Design. https://www.hks.harvard.edu/publications/tough-call-understanding-barriers-and-impacts-womens-mobile-phone-adoption-india#:~:text=Todaypercent20inpercent20Indiapercent2Cpercent2067percent25percent20percent,gapspercent20inpercent20accesspercent20topercent20technology

Bateman, N., and Ross, M. 2020. Why has COVID-19 been especially harmful for working women? *The Brookings Gender Equality Series.*

Bhardwaj, N. 2022. Women and work in India: trends and analysis. India Briefing. https://www.india-briefing.com/news/women-and-work-in-india-trends-and-analysis-24758.html/

Black Feminsm. 2021. What is feminist standpoint theory? https://blackfeminisms.com/standpoint-theory/

Brennen, J.S., and D. Kreiss. 2016. Digitalization. In *The International Encyclopedia of Communication Theory and Philosophy, Volume 1.* Hoboken, NJ, USA: Wiley.

Cameron, J.J., and D.A. Stinson. 2019. Gender (mis)measurement: Guidelines for respecting gender diversity in psychological research. *Soc Personal Psychol Compass* 13: e12506. https://doi.org/10.1111/spc3.12506.

Centre for Economic Data and Analysis (CEDA). 2021. Picture This: Covid vaccination leaving women behind. https://ceda.ashoka.edu.in/picture-this-covid-vaccination-leaving-women-behind/

Chouhan, A. 2020. Structural and Discretionary Bias: Appointment of Female Judges in India. *The Georgetown Journal of Gender and The Law*, 23 (3).

Citi Group and Oxford Matin School. 2016. "Technology at Work v2.0. The future is not what it used to be. *Citi GPS: Global Perspectives & Solutions.* https://www.oxfordmartin.ox.ac.uk/downloads/reports/Citi_GPS_Technology_Work_2.pdf

Clapp, P. 2023. The disproportionate impact of the COVID19 Pandemic on Women in the Workforce. LED Partnership, February 14, 2023

Dalberg. 2021. The disproportionate impact of COVID-19 on Women in India. https://dalberg.com/our-ideas/the-disproportionate-impact-of-covid-19-on-women-in-india/

De Alba, Jaime Moll and Todorov, Valentin. 2022. Measuring and benchmarking the green industrial performance of countries and economies: the GIP index. *Inclusive and Sustainable Industrial Development Working Paper Series,* Working Series 1. https://open.unido.org/api/documents/25133359/download/UNIDO-Publication-2022-25133359

Desai, S., N. Deshmukh, and S. Pramanik. 2021. Precarity in a time of uncertainty: gendered employment patterns during the Covid-19 lockdown in India. *Feminist Economy.* 27 (1-2): 152–172. https://socy.umd.edu/sites/socy.umd.edu/files/users/sdesai/Desai percent20Deshmukh percent20and percent20Pramanik percent202021 percent20Feminist percent20Economics.pdf.

Doss, C., and Kieran, C. 2013. Standards for collection sex-disaggregated data for gender analysis: a guide for CGIAR researchers". CGIAR Gender and Agriculture Research Network. https://www.pim.cgiar.org/files/2012/05/Standards-for-Collecting-Sex-Disaggregated-Data-for-Gender-Analysis.pdf

European Institute for Gender Equality (EIGE). 2019. Gender statistics and indicators. https://eige.europa.eu/gender-mainstreaming/methods-tools/gender-statistics-indicators

European Commission, Emissions Database for Global Atmospheric Research (EDGAR). 2021. GHG emissions of all world countries. https://edgar.jrc.ec.europa.eu/report_2021

Farrell, D., and F. Greig. 2017. The online platform economy: has growth peaked? *SSRN.* https://doi.org/10.2139/ssrn.2911194.

Fry, R., Kennedy, B. and Funk, C. 2021. STEM jobs see uneven progress in increasing gender, racial and ethnic diversity. Pew Research Center. https://www.pewresearch.org/science/2021/04/01/stem-jobs-see-uneven-progress-in-increasing-gender-racial-and-ethnic-diversity/

Gaus, Alexander and Hoxtell, Wade. 2019. Automation and the future of work in Sub-Saharan Africa. Automation and the future of work in Sub-Saharan Africa. Konrad-Adenauer-Stiftung. https://www.gppi.net/media/Automation-and-the-Future-of-Work-in-Sub-Saharan-Africa.pdf

Geldof, M. 2011. Earphones are not for women: gendered ICT use among youths in Ethiopia and Malawi. *Information Technologies & International*

Development 7 (4): 69–80. https://itidjournal.org/index.php/itid/article/download/792/792-2214-1-PB.pdf.

George, Gerard, Ryan K. Merrill, and Simon J.D. Schillebeeckx. 2021. Digital sustainability and entrepreneurship: how digital innovations are helping tackle climate change and sustainable development. *Entrepreneurship Theory and Practice* 45 (5): 999–1027. https://doi.org/10.1177/1042258719899425.

Gilbert, P. 2021. Only 20 percent of Ethiopians are online. Connecting Africa. https://www.connectingafrica.com/author.asp?section_id=761&doc_id=768066

Gurung, L. 2020. Feminist standpoint theory: conceptualization and utility. *Dhaulagiri Journal of Sociology and Anthropology* 14: 106–115. https://doi.org/10.3126/dsaj.v14i0.27357.

GSMA Connected Women. 2020. The mobile gender gap report 2020. https://www.gsma.com/mobilefordevelopment/wp-content/uploads/2020/05/GSMA-The-Mobile-Gender-Gap-Report-2020.pdf

Halpern, M. 2019. Feminist standpoint theory and science communication. *JCOM* 4 (18): C02. https://doi.org/10.22323/2.18040302

International Development Research Centre (IDR). 2020. Women's economic empowerment in Ethiopia https://idl-bnc-idrc.dspacedirect.org/bitstream/handle/10625/59668/IDL-59668.pdf?sequence=2&isAllowed=y

International Monetary Fund (IMF). 2018. The federal democratic republic of Ethiopia selected issues. *International Monetary Fund (IMF)*, https://www.imf.org/-/media/Files/Publications/CR/2018/cr18355.ashx

Jayan, TV. 2020.'Need to reduce gender gap in STEM jobs in India.' The Hindu businessLine. December 14, 2020 New Delhi. Retrieved from: https://www.thehindubusinessline.com/news/science/need-to-reduce-gender-gap-in-stem-jobs-in-india/article33324409.ece.

Jiggins, J. 1986. Women and seasonality: Coping with crisis and calamity. *IDS Bulletin* 17(3): 9–18. https://doi.org/10.1111/idsb.1986.17.issue-3 10.1111/j.1759-5436.1986.mp17003003.x

Kappler, L. 2019. Mobile phones are key to economic development. Are women missing out? *Brookings.* https://www.brookings.edu/blog/future-development/2019/04/10/mobile-phones-are-key-to-economic-development-are-women-missing-out/

Khazan, O. 2018. The more gender equality, the fewer women in STEM." *The Atlantic.* https://www.theatlantic.com/science/archive/2018/02/the-more-gender-equality-the-fewer-women-in-stem/553592/

LeFevre, A.E., et al. 2020. Does women's mobile phone ownership matter for health? Evidence from 15 countries. *BMJ Global Health.* 5 (5): e002524. https://doi.org/10.1136/Bmjgh-2020-002524.

Lennartz, P. 2023. Bridging the Global Gender Gap in STEM. *CEOWORLD* https://ceoworld.biz/2023/01/25/bridging-the-global-gender-gap-in-stem/#:~:text=Consider percent20India percent2C percent20where percent20women percent20comprise,and percent2038 percent25 percent2C percent20respectively.

Lozano, Kevin. 2019. Can the internet survive climate change? How a warming world is sparking calls for a greener web. *The New Republic*. https://newrepublic.com/article/155993/can-internet-survive-climate-change

Mulema, A.A., and Damtew, E. 2016. Gender-based constraints and opportunities to agricultural intensification in Ethiopia: a systematic review. *ILRI Project Report.*

MacQuarrie, Kerry L.D, Edmeades, Jeffrey and Rosenberg, Rebecca. 2022. The gender digital divide: evidence from demographic and health surveys. DHS Analytical Studies. DHS Analytical Studies No. 83. https://dhsprogram.com/pubs/pdf/AS83/AS83.pdf

McKinsey Global Institute. 2019. The future of work in America People and places, today and tomorrow. https://www.mckinsey.com/~/media/mckinsey/featuredpercent20insights/futurepercent20ofpercent20organizations/thepercent20futurepercent20ofpercent20workpercent20inpercent20americapercent20peoplepercent20andpercent20placespercent20todaypercent20andpercent20tomorrow/mgi-the-future-of-work-in-america-report-july-2019.ashx

Melak, A., and S. Singh. 2021. Women's participation and factors affecting their academic performance in engineering and technology education: a study of Ethiopia. *Sustainability* 4 (13): 2246. https://doi.org/10.3390/su13042246.

Mitali, Nikore and Uppadhayay, Ishita. 2021. "India's gendered digital divide: How the absence of digital access is leaving women behind." Observer Research Foundation (ORF) - Digital Frontiers. https://www.orfonline.org/expert-speak/indias-gendered-digital-divide/

New Business Ethiopia. 2022. "IGF calls for promotion of women representation in ICT." https://newbusinessethiopia.com/technology/igf-calls-for-promotion-of-women-representation-in-ict/

Organisation for Economic Co-operation and Development (OECD). 2018. Bridging the digital gender divide. Include, Upskill, Innovate. *OECD*, https://www.oecd.org/digital/bridging-the-digital-gender-divide.pdf

OECD. 2021. The inequalities-environment nexus: towards a people-centered green transition. *OECD Green Growth Papers*, no. 2021/01. https://www.oecd-ilibrary.org/environment/the-inequalities-environment-nexus_ca9d8479-en

Pennington, James. 2022. 3 ways the circular economy is vital for the energy transition. ClimateChampions. https://climatechampions.unfccc.int/3-ways-the-circular-economy-is-vital-for-the-energy-transition/?gclid=Cj0KCQiA_bieBhDSARIsADU4zLdXqknvL3XRViBrggWZCLI6GGiKBpSfzL8hEqWjRO7l0_vPSGF5BhYaAvFxEALw_wcB

Singh, B., et al. 2019. Industry 4.0 and future of work in India. *FIIB Business Review* 8(1): 9–16. https://doi.org/10.1177/2319714519830489

Singh, A., J. Svenssona, and A. Kalyanpur. 2010. The state of sex-disaggregated data for assessing the impact of climate change. *ELSEVIER Procedia Environmental Sciences* 1: 395–404. https://doi.org/10.1016/j.proenv.2010.09.027.

Stoet, G., and D.C. Geary. 2018. The gender-equality paradox in science, technology, engineering, and mathematics education. *Sage Journals* 29 (4): 581–593. https://doi.org/10.1177/0956797617741719.

Pew Research Center. 2021. Internet/Broadband Fact Sheet. https://www.pewresearch.org/internet/fact-sheet/internet-broadband/#panel-9a15d0d3-3bff-4e9e-a329-6e328bc7bcce. Accessed 20 Jan 2023

Tara Satyavathi, C., et al. 2010. Role of farm women in agriculture. *Gender Technology and Development* 14 (3): 441–449. https://doi.org/10.1177/097185241001400308

The Beautiful Truth. 2022. What is the green transition? https://thebeautifultruth.org/the-basics/what-is-the-green-transition/

UN Women. 2020. Intersectional feminism: What it means and why it matters right now. https://www.unwomen.org/en/news/stories/2020/6/explainer-intersectional-feminism-what-it-means-and-why-it-matters. Accessed 9 Dec 2021

UN, Department of Economic and Social Affairs, Statistics Division. 2016. Integrating a gender perspective into statistics. *Studies in Methods*, Series F No. 111. New York: UN. https://unstats.un.org/unsd/demographic-social/Standards-and-Methods/files/Handbooks/gender/Integrating-a-Gender-Perspective-into-Statistics-E.pdf

UNDP. 2020. COVID-19 Global Gender Response Tracker. Retrieved from https://www.undp.org/publications/covid-19-global-gender-response-tracker-fact-sheets

U.S Bureau of Labor Statistics (BLS). 2021. Women in the labor force: A databook. *U.S Bureau of Labor Statistics*, https://www.bls.gov/opub/reports/womens-earnings/2019/home.htm#:~:text=In%202019%2C%20median%20weekly%20earnings,54%2C%20with%20earnings%20of%20%24904

U.S Bureau of Labour Statistics (BLS). 2022. *Employment projections: 2021–2031 summary*. Washington, DC: U.S. Bureau of Labor Statistics. https://www.bls.gov/news.release/ecopro.nr0.htm

Wieser, C., Menaal, E., and Tefera, G.E. 2020. Assessing Ethiopian women's vulnerability to the COVID-19 pandemic, December 07. https://blogs.worldbank.org/africacan/assessing-ethiopian-womens-vulnerability-covid-19-pandemic#:~:text=Although percent20worker percent20layoffs percent20were percent20limited,off percent20in percent20June percent20were percent20women.

World Bank Live. 2021. Key green transitions: how systems are changing for people and the planet. https://live.worldbank.org/key-green-transitions-climate-systems

World Economic Forum (WEF). 2022. Global gender gap report 2022. https://www3.weforum.org/docs/WEF_GGGR_2022.pdf

Yardi, P., M. Marathe, and K. Toyama. 2017. Differences in STEM Gender Disparity between India and the United States. HCI Across Borders Symposium (HCIxB 2017) at the 2017 CHI Conference on Human Factors in Computing Systems. http://www.hcixb.org/papers_2017/hcixb17-final-37.pdf

CHAPTER 13

Concluding Chapter: Reimagining a Green and Digital World

Kristie Drucza and Amira Kaddour

13.1 Introduction

The SDGs challenge us to find new ways to build a sustainable future. This chapter uses utopian theory and Rawls's theory of justice to summarize the 11 preceding chapters, with the hope of stimulating policymakers and scholars to creatively embrace equality and diversity as the *keys* to building resilient and sustainable ecosystems. Intersectionality and inclusion will be critical for such transitions since diverse decision-making processes make solutions more sustainable and equitable.

K. Drucza
Includovate, Kampala, Uganda
e-mail: kristie.drucza@includovate.com

A. Kaddour (✉)
National School of Advances Sciences and Technologies, University of Carthage, Amilcar, Tunisia

Includovate, Tunis, Tunisia
e-mail: amira.kaddour@enstab.ucar.tn; amira.kaddour@includovate.com

K. Drucza et al. (eds.), *Centering Gender in the Era of Digital and Green Transition*,
https://doi.org/10.1007/978-3-031-38211-6_13

Science fiction explores dystopian futures and often depicts individuals abusing their power and greedily holding on to scarce natural resources. This book, in contrast, provides positive evidence to argue for "a different, better way of being" (Levitas 2011), with the different authors exploring ways to create more equitable futures. Levitas argues that "the transformation of reality and the realization of utopia depend … upon not only wishful thinking but will-full action." We hope this book will stimulate people's desire to act for an equitable and sustainable future, and points to the way we can best go about this.

13.2 Summary

The **first part of this book** discusses progress toward the SDGs and highlights how more investments are needed. For Rawls (1971) social justice entails institutions and societies operating for everyone in a fair and equitable manner, free from prejudice and bias. Rawl's theory of justice focuses intently on the least well-off individuals, as do the SDGs in their determination to "leave no one behind." For Rawl, though, the ideal society can have some inequalities (in wealth, power or consumption) so long as those inequalities can benefit the disadvantaged. However, the previous 12 chapters reveal that when the basic structures of society are not just, they cannot protect or benefit the least advantaged.

In the second chapter, Joshi, Mensah, and Correa highlight that gender equality is a fundamental human right and explain why gender equality is crucial to meaningful progress across all 17 of the SDGs. The authors view intersectionality as a necessary lens for understanding and achieving gender equality and suggest how we can better progress equality across all the SDGs, especially those that relate to green and digital transitions. Given the limits of current data, they argue that states should bolster gender-specific targets and indicators for the SDGs, and that new sex-disaggregated and ICT-related indicators might best be incorporated in routine national censuses.

In **chapter three** Kaddour and Ghbara discuss the relationship between gender inequalities, poverty, and disparities in the era of digital transformations and green transition. They argue that if development strategies do not properly consider inequalities, they will reinforce poverty and undermine development. It is not—the authors posit—only macro-regional data that is required, but country-specific data (on gender, race, age, disability, etc.).

In **chapter four**, Kaddour, Louhichi, and Fourati explain the way climate change impacts women differently at the environmental, economic, and social levels. Since women bear a disproportionate burden of climate change impacts, the UNFCCC has taken steps to include gender dimensions in the Paris Agreement, the Lima Work Program, and the Gender Action Plan. However, more action is needed to raise awareness and promote women's participation in politics, technology, and science initiatives that promote climate adaptations and mitigations.

The **first part** of this book argues that we lack critical data and targets and must address these urgently, and as the first step toward addressing inequalities related to the digital and green transitions. While there are some positive trends, the urgency of the climate crisis and the slowness of progress to date currently paint a bleak picture and suggest that without escalating efforts we will not reduce gender inequality to achieve the SDGs.

Part two of the book presents its vision for three strategic sectors: agriculture, energy, and transport, explaining why these sectors are particularly crucial for women in terms of green and digital transitions. These chapters demonstrate that where you are born determines the quality of services you receive and (often) limits your opportunities (this is in contrast to Rawls's idea of using rules to create a fair society).

Lefor and Ganguly in **chapter five** discuss how gendered dimensions are often not fully considered when it comes to project implementation and research, especially in climate change and agriculture. The chapter combines a desk review with the experiences of 20 gender experts in climate change and agriculture from across Asia and Africa. It argues that climate change has a disproportionate impact on women and girls, and that failures to recognize or act on this reality can halt or even reverse progress toward gender equality. Although, encouragingly, gender is receiving more attention at the program level and overall funding for it has increased, budgetary allocations for specific gender equality programs are currently lagging behind. In addition, programs that do integrate gender often do so in a tokenistic way that only engenders superficial results, and do not define and detail the critical pathways that can create systematic global long-term changes.

In **chapter six**, Curbelo and Haile highlight the importance of gender-responsive national energy policies, which aim to ensure inclusive access to affordable, reliable, sustainable, and modern energy as stipulated in SDG 7. The research shows that gender-responsive policies are positively correlated with the proportion of women employed in the industry and service sectors, but negatively correlated with the proportion of women employed in the agriculture sector. The findings provide empirical evidence of the beneficial consequences of gender-responsive national energy policies, especially in terms of improved macroeconomic outcomes for East and Southern African countries.

In **chapter seven**, Talevska, Kaddour, and Burlacu discuss gender issues in relation to transportation, mobility, and road safety. The authors point out that discriminatory social norms currently limit women's access to the transport sector, impede their mobility, and put them at risk. They suggest giving a new focus to transportation and its contribution to gender (in) equality, including disaggregating data, measuring gendered road safety issues, and supporting inclusive transport systems.

The **third section of the book** argues that merely including women into existing unequal systems is not enough to achieve gender equality. Strengthening education, female leadership and combating new forms of violence are also needed for women to effectively participate in economic and social life. Rawls suggests that in a just society, rules will be fair because each decision-maker will contemplate the common good. This chapter, however, suggests that rules alone are not enough.

In **chapter eight**, Laabidi and Gtifa posit that women can—and should—take leadership roles in terms of promoting sustainable consumption and production, preserving the environment, and overseeing resource allocation in households and communities. Improving their status, accessing to environmental resources, and educating them in technology and entrepreneurship are necessary to unlock their full potential. This requires governments to initiate programs that provide financial support, gender-responsive procurement, and address the financial challenges faced by women.

In **chapter nine**, Kaddour, Gmach, and Elghoul consider the gendered dimensions of the STEM field, showing how failures to mainstream gender issues can further increase digital divides and exacerbate current gender inequities. Closing gendered inequities in STEM requires a multi-faceted approach including encouraging diversity in the tech sector, supporting girls and women to access STEM education, and advocating for tailored technologies for women.

In **chapter ten**, Maazouz and Kaddour present how digitalization can foster new forms of GBV, but can also help address and prevent GBV. Tech designers and developers must take a proactive approach to inclusively involve women at all stages, ensure their active participation, incorporate preventive measures and ethical considerations, and foster technology that is accessible and safe for all, including those from marginalized communities.

While none of these chapters suggest that only women can solve women's problems, they do suggest that decision-making and policymaking need to be more equitable. In line with Rawls, they argue that laws, policies, and rules with important social outcomes need to be collaboratively developed, and that the participation of women and other marginalized groups is essential to ensuring fair and just transitions.

The **fourth part of the book** focuses on monitoring economic developments that foster just transitions. For Rawls, monitoring the process of rule-making and its implementation is more important than monitoring outcomes. This section of the book argues that globally agreed goals (i.e., expanded SDGs) need to be set, with active data collection, analysis, and monitoring to ensure new forms of inequities are not created and that some groups do not benefit at the expense of others.

Ganguly and Nikolova, in the **eleventh chapter**, argue that gender equality and economic development are closely intertwined and that gender aspects should be an essential consideration for economic development. Employment, social protection, unpaid work, and the care economy are the focus sectors discussed. The chapter points out that the post-COVID-19 recovery period can provide a window of opportunity to strengthen current economic models by ensuring they better include gender issues and embrace the green economy. Pandemic-related economic measures can help support the investments needed for structural transformations as well as the technological innovations needed for sustainable development.

In **chapter twelve** Drucza and Curbelo use feminist standpoint theory and intersectionality to frame the current data and indicators that relate to digital and green transitions. The chapter shows that across three countries—Ethiopia (low-income), India (middle-income), and the USA (high-income)—'green' data is less likely to be sex-disaggregated than digital data. It argues that inequalities will not simply and evenly reduce as wealth is created, and that each country will require a tailored approach to ensure equities are fostered during digital and green transition. This will require understanding the needs, positions, and experiences of different groups, gender analyzing any data generated, and ensuring effective programs and initiatives are informed by this information.

13.3 Conclusion and Future Research Agenda

The world is rapidly evolving. If we are to create sustainable and inclusive societies, we need to create cultures that respect the needs of others and systems of governance that ensure equal opportunities, and build economic systems that allow equal access, participation, and benefits for all groups of people. Reimagining power and control is necessary to ensure particular groups are no longer able to impose their values on others because of their privilege, power, and/or status in society.

With the deadline for the SDGs only seven years away (2030), urgent action and investments are needed for women and girls. The turbulence caused by COVID-19 has worsened gendered outlooks in many countries, and we need to work together as a global community to comprehend and combat the discrimination and marginalization that lead to social inequalities and injustices. To facilitate this, policymakers must mainstream gender in socioeconomic policies and national data collection, and capture and analysis must become more sophisticated in terms of analyzing and quantifying macro- and micro-inequalities.

This book is limited by a research base predominately created, informed, and biased toward men. We propose a future research agenda that is solution-focused and can start by considering the following questions:

- How exactly are gender and other forms of inequality created, maintained, and reproduced in green and digital practices at the household, community, market, and state levels?
- How are patriarchy and harmful masculine norms perpetuated through digital media and the green industries? How might we transform these into more positive and equitable norms?
- How do women and other marginalized groups (LGBTIQA+, persons with disabilities and migrants) use digital technologies differently? What may they need for safer solutions?
- Which green and digital technologies are preferred by women in different sectors (agriculture, energy, transport etc.)? How might these be utilized to promote climate adaptations and mitigations?
- How do power relations within green and digital industries maintain the status quo?
- How can we influence gender decision-makers to become more equal in their own lives as a means of making policies more gender equal?

- How can female leaders promote sustainable consumption and production practices at the community and household levels?
- How exactly can gender-responsive national energy policies improve macroeconomic outcomes?
- What budgetary allocations are needed for which types of gender-equal solutions?
- Did any pandemic-related economic measures improve women's unpaid work burden?
- Which kinds of affirmative actions (quotas, targets, scholarships, gender-responsive procurement etc.) can best promote gender equality in STEM? And in SMEs across the green and digital industries?
- Which digital and green indicators are most needed in national censuses to understand if women are being left behind by transitions?
- What should we be measuring to ensure new forms of inequities are not created and that some groups do not benefit at the expense of others during technological transitions?

The authors encourage their readers to pursue these questions, and to bolster the research base with more diverse perspectives.

There is no "universal utopia" (Levitas 2011), but a sustainable future is possible. This book has highlighted key steps to creating that, showing how we can rethink social and economic models, to equally recognize and value all people. It has also shown how disaggregating data and adopting an intersectional lens can make the digital and green transitions more inclusive, equitable, and sustainable. We hope this book inspires our audience to see equality and diversity as the keys to building resilient and sustainable ecosystems.

References

Levitas, R. 2011. *The concept of utopia*. Oxford: Peter Lang International Academic Publishers.

Rawls, J. 1971. *A theory of justice*. Oxford: Oxford University Press. 1999.

Index[1]

[1] Note: Page numbers followed by 'n' refer to notes.

K. Drucza et al. (eds.), *Centering Gender in the Era of Digital and Green Transition*,
https://doi.org/10.1007/978-3-031-38211-6

The manufacturer's authorised representative in the EU is Springer
Nature Customer Service Centre GmbH, Europaplatz 3, 69115 Heidelberg,
Germany. If you have any concerns regarding our products, please
contact ProductSafety@springernature.com

Printed and bound by CPI Group (UK) Ltd, Croydon, CR0 4YY
28/09/2026
02224072-0002